RAISED BY THE ZOO

Gerry Creighton was operations manager and elephant keeper at Dublin Zoo for thirty-six years and the face of RTÉ's popular series *The Zoo*. He is now a public speaker and consults for zoos and wildlife parks around the world, promoting optimal care and wellness for elephants in human care. He lives in Dublin with his family.

RAISED BY THE ZOO

MY LIFE WITH ELEPHANTS AND OTHER ANIMALS

GERRY CREIGHTON

GILL BOOKS

Gill Books
Hume Avenue
Park West
Dublin 12
www.gillbooks.ie

Gill Books is an imprint of M.H. Gill and Co.

© Gerry Creighton 2023

978 07171 97514

Designed by Bartek Janczak
Typeset by Typo•glyphix, Burton-on-Trent, DE14 3HE
Edited by Louise Ní Chríodáin
Copy edited by Sylvia Tombesi-Walton
Proofread by Liza Costello
Printed and bound in Great Britain by Clays Ltd, Elcograf S.p.A.
This book is typeset in Adobe Garamond Pro 12pt.

For permission to reproduce artwork, the author and
publisher gratefully acknowledge the following:
© Shutterstock/HN Works

The paper used in this book comes from the wood pulp of sustainably managed forests.

A CIP catalogue record for this book is available from the British Library.

5 4 3 2 1

This book is dedicated to my wife, Leona,
and my children, Mia and Zac.

CONTENTS

ONE

THE ELEPHANT
IN THE ROOM

This is the story of a boy from Dublin's inner city who found himself inspired by elephants and set out on a mission to change how humans treat them across the globe.

My father worked as a zookeeper for over half a century, and he often reflects on how much positive change has happened in the last few decades. Although I'm passionate about zoos, their important role in education and in helping endangered species, I also know the problematic origins of menageries and animal collections that were developed for display purposes only. Zoos have evolved dramatically and still have much to do. This book is part memoir and part manifesto about the 'elephant in the room' – that is, how we need to relegate the mistreatment of elephants and other animals to the past and reimagine our future with them.

Elephants are extremely powerful yet agile animals, and their body is a marvel. Take their extraordinary trunk, for example. A fusion of the nose and the upper lip, the trunk has more than 100,000 muscle units. It bears a resemblance to the human tongue – both are what is known as a muscular hydrostat (the octopus tentacle is another example). Elephants can use this structure to pick up something as small as a needle or as large as a tree trunk. They use it to breathe, drink, eat, smell, wash, snorkel and communicate.

But it is their minds that make them so beguiling. Elephants are sentient, smart and enthralling to work with. They have always held a fascination for humans, and we love them for many different reasons. Yet, although we are in awe, we have not been kind to elephants. Historically, we have exploited their intelligence, their sociability and their cooperative natures – and this exploitation continues, particularly in the home-range countries. Because of their charismatic appeal, we were happy to keep them close under any conditions. We used violence and dominance on them so we could stand beside them, feel that we had dominated them, or have them work for us by carrying us and our loads as beasts of burden or perform for our amusement.

Their history in Dublin Zoo, a place I have known all my life and worked at for 36 years, is no different. The archives contain photographs of elephant rides, elephant calves on their back legs performing circus-like tricks, and elephants surrounded by people. They look like fun pictures, but these are elephants that were probably taken from the wild, removed as neonatal calves from their mothers and family units. The impact of this sort of trauma on these animals was simply devastating. Elephant society is based on learned behaviour throughout a lifetime. These animals learn from each other, and no human can teach them anything about being an elephant. When you separate a creature with an emotional intelligence similar to ours from its family and from other elephants, huge trauma ensues.

Young elephants were shipped off on boats to Europe, to be beaten into submission until they did what was demanded of them. Because of an elephant's size and strength, zookeepers saw fear as the only way to control these majestic animals, to keep themselves and the public safe.

The way elephants were managed in the past – and, unfortunately, the way they are still managed in some parts of the world – is called free contact. This method requires keepers or trainers to share the same space with the elephant and use a hook or an ankus to dominate it with fear. If the elephant does not perform as the human wants it to, it is prodded in a sensitive part of the body – soft areas like behind the ear, the temple or behind the knee. When I began as a keeper, the elephants in our care had already had their spirits broken. The mere sight of a hook was enough to ensure compliance.

Protected contact – where elephants and humans never share the same space – was first developed in zoos for keeper safety, but it has transformed elephant welfare too. Combined with positive reinforcement training (where an animal is rewarded rather than punished), it has changed how we care for these wonderful creatures.

It has been almost 20 years since Dublin Zoo left free contact where it belongs, in history, and embraced a new approach to animal care. We built a protected-contact wall to stop us from invading the elephants' domain. Keepers (or animal carers, as they're now more commonly known) now remain on one side of this wall and elephants on the other. We took an elephant who had been dangerous in free contact and watched her preside over multiple births and lead a multigenerational herd. Our only contact was during positive reinforcement training, to allow us veterinary access via ports or openings in this wall. Even a six-tonne bull happily allowed us to pare its nails, take a blood sample and spend hours in a transport crate, without sedation.

No animal in human care should have a consequence of fear or pain for being in our world. However, there are still people in zoos across Europe, North America, Asia and India who believe that free contact (that is, using the hook) is the best way to manage elephants. I've been in the same room as these people, who think that it is okay to use force and inflict pain to control an elephant, to show it that they are 'the boss'. I have given presentations to groups, including the Elephant Managers Association in the US, where people sneer at 'the guy over from Ireland' who trains elephants with kindness. I start by saying: 'I did the same as you guys. I had a hook. Thankfully, I never had to use it, but the reason why I never had to use it – and you may never have to use it – is that the shit was already beaten out of these elephants, and they were bankrupt and fearful. They were afraid of what I might do to them.' And some of these guys slink back into their chairs a little.

Even from a young age, when I looked at the Elephant House and enclosure, I'd think, 'There's got to be a better way.' When I joined the zoo, our Elephant House was from the 1950s, designed when the criteria were containment, confinement, safety and ease of cleaning. None of the biological reference points of these incredibly smart creatures was understood. Elephants were kept without the physical or mental stimulation required for such an intelligent animal, and this was one of the things that really concerned me, and so many other keepers.

Keeping elephants on hard, barren concrete surfaces, where they stand in their own urine and faeces, has huge consequences for their welfare and their feet. There's a myth that elephants can sleep standing up. That's rubbish – they can rest or lie against something, but they can't sleep properly. There are elephants in human care that have never been able to lie down because their only choice is concrete. This is horrific. We have to look to nature for inspiration. Elephants in the wild have the answers for better-designed habitats. But this is only one of the issues that need to be addressed for elephants in human care.

Elephants in the wild also need our attention. Every 15 minutes, an African elephant is killed and butchered. For every few pages that you turn in this book, an elephant will be destroyed. There's still huge demand for ivory in countries all over the world. In both Asia and Africa, many elephants are no longer forming tusks. This is an adaptation that may help them to survive – nature is trying to save them. However, they are also being harvested for leather and meat. It's an ongoing battle to try and educate people.

But it's not just the people who live alongside elephants that need educating. The mentality of someone who would hunt an elephant for sport, paying up to $100,000 to put a bullet in an elephant's head, is completely unfathomable to me. I will never understand why a person would travel to an African reserve to drop an elephant to the ground with a big-game rifle, destroying and decimating a family unit, just to get a photo to display on their wall.

There are two genera of elephant (African and Indian), with three species altogether, one of which, the Asian elephant, has four subspecies. African elephants are the world's largest land mammals, and Asian elephants are the second largest. All are endangered, and all are keystone species, meaning that they play a critical role in our ecosystem. If elephants survive and thrive in their environment, everything thrives. However, there are thought to be fewer than 30,000 Asian elephants left in the world, and only about 400,000 African elephants. Asian elephants can be found in forested regions of India and throughout Southeast Asia, including Myanmar, Thailand, Cambodia and Laos. There are only 250 wild elephants left in China, but I've managed to see some of them.

Dublin Zoo is part of the European breeding programme for the Asian elephant, part of the effort to protect them from extinction. The main threat comes from deforestation for agriculture and housing. Agriculture has also led to increased conflict between elephants and humans. As elephants look for more space, they get killed by farmers for entering their land and eating their crops. They are often viewed as pests, particularly in India, where they come into tea plantations and destroy the irrigation. They don't actually eat the tea plants, but they rip up the pipes to find water. Other threats to elephants include poaching and being captured for the tourist trade.

When people ride an elephant in Thailand, they need to understand that for an elephant to accept being ridden, it has been threatened and abused to the point of compliance. Its spirit has been crushed. They look like elephants, they are grey and have trunks, but emotionally, they are bankrupt. The ankus hook – used to hit pressure spots behind the ear, genital region or inside leg – has been replaced by less obvious small sharp pins or nails. In my eyes, riding an elephant makes you an accessory to that abuse.

Culture is also used as an excuse for abuse in many of the elephants' home-range countries, like India, Thailand, Indonesia and China. People, including tourists, flock to see India's temple elephants – majestic bulls with bleached white tusks in ceremonial headwear who have been shackled, chained and beaten into submission so that they will stand still for hours. Away from the temples and the parades, they may be left chained in blank, concrete cells. These are the same creatures that are programmed by nature to spend 18 hours of their day foraging and feeding.

In their natural habitats, female elephants are highly sociable and live in close-knit, multigenerational family herds. The matriarch leads and takes responsibility for, the herd. She's not necessarily the biggest or the strongest elephant, but she's the smartest one, the one that has the information, the retention. Elephants are known for their excellent memory, and matriarchs will remember ancient seasonal migration routes and watering holes, leading the herd through these routes every year. The matriarch controls the herd in times of panic and lends support to young females giving birth for the first time, offering vocal encouragement, as well as touching and holding them.

These are all traits and behaviours that we once thought were exclusively human. These intelligent animals have a highly evolved neocortex, similar to humans, great apes and some dolphin species, and they display many traits in common with us, including compassion, grief and altruism. When a herd member dies, elephants may come back year after year to where the body lies. Herd members will share food and socialise, and even care for each other's offspring. Herds that know each other will stop to touch and talk when they meet in the wild.

Elephants are very communicative, but their communication happens in many different ways, including touch, smell, body language and vocalisations. If Dublin's elephant herd come over to me, they will make childlike twirps as they run – their greeting sound. Often, when they are getting washed or rewarded with food during training, they make a contented rumbling sound, a bit like a low drum roll.

We have learned that elephants have unique family sounds handed down from one generation to the next. To talk to each other, they also use a frequency, an infrasound that humans cannot hear, which can travel for over a mile. They can hear other ultrasounds, too. Before the Indian Ocean tsunami in 2004, elephants were witnessed shrieking and running for higher ground.

Smell is of huge importance to elephants. If the trunk is up in the air, in an S-hook shape and the nostrils are pointing in one direction, they are capturing scents. Elephants have over 2,000 olfactory receptors in their brain, more than that of any other animal we know. They can smell water from miles away, and one study showed that they can also smell TNT, the main ingredient in landmines. It's no surprise that they use urine as a messaging system. A group of 20 or 30 elephants might spread over quite a distance to find resources when they are feeding across the African savanna. To keep a check on the herd members, the matriarch uses urine to track them. She may not pay any attention to urine from other elephants but will check the urine trail of her own herd. She'll know from the urine how long it's been there, because it contains proteins and chemicals that fade over time and, through her experience, she knows when they have passed through. It's effectively like leaving a timestamp of where they have been.

An elephant's massive brain has the ability to learn throughout the animal's life. Elephants are constantly acquiring new skills, new ways of interacting and new ways of communicating. There's a lot we can learn from them.

TWO

BEWARE OF
THE LEOPARD

I'm a second-generation zookeeper. My father Gerry Snr – Da – was a keeper at Dublin Zoo for 52 years. Like a lot of local lads from around Stoneybatter, he got summer work as a pony boy. He was 14 when he wandered into the Phoenix Park and through the back gate of the Zoological Gardens to ask about a job. When he walked through the yard, all the ponies began to follow him, and as he smelled the horses and the hay, he thought, 'This is where I'm going to spend the rest of my life.'

In those days, it was working with the ponies that identified someone with the ability to progress to become a keeper – and Da had that ability. He can still recite all their names: Blackberry, May Blossom, Tiny, Silvermist, Little Ginger, Big Ginger, Bubbles and

Dot. Although he's long retired, he still dreams about them – and about the zoo.

My grandfather had wanted him to join Guinness, where he and my uncle Tommy worked. The brewery and Córas Iompair Éireann (CIÉ) were seen as the great local employers, with lots of benefits. Da, however, was determined to become a keeper.

In 1958, when he started, there was only one building on the far side of the zoo lake, the polar bear pit and some wild wallabies. The rest was a jungle of trees and plants. Before his time, they held swimming and high-diving championships in the lake – there are pictures of Johnny Weissmuller, who starred in the *Tarzan* movies, diving in. Da and the other young keepers swam there too. I was told there had also been fishing competitions, but the fish were killed by the runoff from coal stored along the park's roadways during World War I. The lake also became the divider when staff were split into two teams: the near-side team, who worked near the offices, and the far-side team, who worked with animals on the far side of the lake. There was a lot of light-hearted slagging between the two sides.

It was Paddy Mahon and Ned 'Heem Haaw', a military man who had been gassed in the First World War, who taught Da that details mattered when looking after animals. Grooming had to be done meticulously: the pony's mane and tail cleaned and brushed, and all particles of hay or straw removed; the teeth checked; the feet washed and cleaned out with a pick, including a check for disease in the sensitive frog of the foot; the hooves painted with oil. Learning how to put on the saddle and bridle without pinching a pony was a lesson Da never forgot.

The zoo then was very much a social place. Under the stewardship of director Terry Murphy, there were lots of parties and events held on the grounds. There was also still a lot of British influence through the Royal Zoological Society. Terry was a charismatic man, but he was from an era when they just bought up animals and filled up the cages. It was a typical Victorian zoo.

Da's first permanent job in the 1960s was in the zoo stores. He delivered tea chests of food to the different houses and enclosures on an electric buggy. The chests that he delivered to the Giraffe House were packed with dog biscuits – the president of the Zoo Council at that time was George Shackleton of Shackleton Mills, which made Ovals biscuits for dogs. According to Da, Dublin was the only zoo in the world that fed its giraffes dog biscuits, and the giraffes were breeding better than in other zoos.

My dad also remembers a lorryload of wild goats that had been brought in to feed the lions and tigers. The goats escaped and ran over to the far side of the lake, which was like a jungle at the time. Da was one of the workers sent over to round them up, including huge billy goats that would charge at them. It took weeks to catch them all. Usually, however, the big cats were fed with donkeys and old horses suffering from a range of complaints, or racehorses that had fallen. They were put down and cut up in the zoo slaughterhouse. When companies stopped using horses for transport, Da recalls all the old gas company horses arriving to be put down, still stinking of gas.

Even then, keepers kept a close eye on what animals liked or disliked, and they noticed if they were off their food. Requests would come in for less or more food or something new: 'Sally the chimp won't eat tomatoes. Can you increase the bananas and leave the tomatoes out?' or 'Harry the great Indian hornbill is not keen on those chopped bananas. Can you give me more grapes?' All the primates in the Monkey House received a plate of warm milk every morning and night with added vitamin drops. The higher order of primates, such as the chimps and orangutans, had their own mugs. Kenny the olive baboon would turn his tin plate sideways and, with deadly accuracy, throw it through the bars at his chosen target. Da soon learned to hone his ducking reflexes.

As a trainee, Da served as a 'floater' (a role I would later fill myself), working in different departments when keepers were sick or on holiday. He might be filling hay nets for the dromedary camels or giving the Monkey House a deep clean. Some of the senior keepers were almost like celebrities at that time – men like Jimmy Kenny, who was in charge of the elephants, camels and hippos, and reptile keeper Tommy Kelly, who would drape snakes around visitors' necks in the Reptile House.

That house included another Tommy, a huge crocodile that was named after his keeper. Da is still baffled by how Mr Kelly managed to move this massive creature on his own into a new tank. A plan had been drawn up – involving experienced keepers, vets, sedation, nets, ladders and planks of wood. However, when the team arrived in the morning to begin the dangerous transfer, the crocodile had already been moved; only a couple of planks remained in the cage. Tommy refused to tell even director Terry Murphy how he had done it, but he later told Da that he had 'hypnotised' the crocodile and walked it to his new tank in a trance.

———————

Da met my mother Catherine in the zoo. She was working as a cashier in the restaurant for the summer. He spotted her walking across the lawn, turned to a friend beside him and announced, 'That's the girl I'm going to marry.' He's still convinced that he can sometimes predict the future. She was 17, and he wasn't much older. He was a good-looking bloke, with his hair gelled back like Elvis, his hero. She always laughed telling us about the stink off him when he appeared in the restaurant to ask her out, holding a bucket of mackerel on his way to feed the sea lions. Apparently, that's why she didn't say yes straight away.

They married a few months before my mother turned 20. My sister Margaret and I are Irish twins – she's 10 months older than me. And then there's Eithne and Cathy. We were all born on Ivar Street, in Dublin's Stoneybatter. James, the youngest, was born after we moved out of the city to Blanchardstown, in West Dublin.

I arrived on 29 February 1968, a leap-year baby … so I'm technically only 14 as I write this. By the time I was born, Da was a lion keeper, much to his delight. He had been fascinated by the cunning and intelligence of lions since reading *The Man-eaters of Tsavo* as a child. This was the story of the lions that terrorised the workers building the Kenya–Uganda railway in 1898, killing an estimated 135 people. He would cut out pictures of animals from the newspapers, and save up his pocket money to write to zoos asking for information and photos of lions and tigers. His desire to work with lions when he was older wasn't even dampened when an escaped lioness mauled a young petrol pump attendant and renowned lion tamer Bill Stephens in nearby Fairview. Stephens had been keeping three lions in an enclosure behind the garage. The other two lions were confiscated and brought to Dublin Zoo. As for Stephens, he was killed by another lion a couple of years later.

Because of Dublin's unique history with the African lion, it was one of the most prestigious keeper jobs. It was a Victorian zoo, with animals in cages and pits. The lions were kept in the Roberts House, which had been built in memory of a president of the Royal Zoological Society of Ireland, and the big cats paced up and down in square boxes behind bars.

Despite the animals' living conditions, Dublin had been renowned for breeding big cats since the late 19th century. There were a lot of theories about this success in producing robust animals. One was that they were fed horse meat. It was thought that because the horses were eating good grass, the nutrition produced healthier, stronger lions. At one stage, selling lions earned the zoo more money than annual subscriptions. They became a very famous export. One went on to become the iconic Metro-Goldwyn-Mayer lion.

I've been around Dublin Zoo and the Phoenix Park since I was a baby. My first memories of it are holding Da's hand as we walked into the Roberts House, and the stench of cat urine when he opened the door. It was like that every morning when the house was opened up to air out for visitors. The ammonia steaming off the wooden cage floors brought tears to my eyes, and I'd hold my breath, waiting for the fresh air to come in and dilute the smell.

The zoo was a very intimidating but exciting place to be as a child. I was always hanging on to the back of Da's heels at the weekends. I pushed the meat trolley around (it was painted like a zebra) as he would hook the meat and feed it to the leopards and the lions. The drama and energy of those ferocious cats had me mesmerised from a young age. Their huge paws would reach out through the bars, and I was fascinated by how they retracted their claws after pulling in the meat. The sound in that house was incredible.

If you're very close to a big cat when it roars, you'll notice how the air actually vibrates. The energy just passes right through you. The voice of any animal, including humans, is produced by air from the lungs flowing past the vocal folds, or cords. In most species, the vocal folds are triangular and protrude into the animal's airway. However, the vocal folds of lions and tigers are different from other animals (and from other big cats – like cheetahs, for example, who can't roar). In lions and tigers, the folds are flat and square, which means they can withstand strong stretching. This makes it easier for the tissues to respond to the passing airflow, allowing louder roars at less lung pressure. The lion can create a sound that carries across the savanna when needed, so in a confined area, the energy and the power are phenomenal.

I don't think I was ever afraid of them. I suppose I fed off my father's confidence – he was a top-class cat keeper. He understood them and how to manage them. It was far more dangerous than it is now, so keepers had to have incredible reactions and judgement. He was taught the way to catch a leopard was by its tail, holding a brush in the other hand. Once Da was cleaning a lion cage when another keeper let its occupant, Rusty, back in by mistake. Both he and the lion got a shock, but Da instinctively instructed him to go 'amach, amach' (he spoke to Rusty in Irish), and Rusty slouched back into the other cage. Another time, he rounded up five lions who had escaped out of their enclosure by rattling his keys – the signal for dinner time – and herding them like he did the flamingos, with his arms open wide. Despite the risks, he has only one scar, which he got when feeding a black panther. My mam came into the house with Margaret and me, and as my dad looked around, the panther lashed out and scratched his hand.

The most skilled keepers were those who could get in and out of the cat houses and monkey houses quickly. Speed proved essential the day Bim, a two-tonne elephant seal, took a fancy to Da, prompting a hasty exit.

Keepers also needed to be fast and fully alert going into the rhesus monkey pit, where there were about 20 monkeys led by a dominant male known as Hitler, who would preside over their cleaning while perched on a flagpole. The keepers would go in through a small door, but getting out was very tricky because Hitler would try and take a lump out of their backsides if they weren't quick enough. Keepers were warned not to look Hitler in the eye or he would charge them.

That particular rhesus monkey troop would later stage an escape, despite an electric barrier. Breaking up into different hostile gangs, they spread throughout the zoo and beyond, looking for food and creating mayhem. Da was one of the keepers who responded to a call from Áras An Uachtaráin, where he netted a monkey clutching a pound of sausages in the presidential kitchen. Others took up residence in a housing estate in Cabra West. Eventually, they were all caught, except an increasingly aggressive Hitler. He continued to raid houses for food for a week, before being shot and killed by zoo director Terry Murphy on a Cabra rooftop.

For many years, two of the highlights of a zoo visit were the Chimpanzee Tea Party on the lawn and the bone-crunching spectacle of feeding time at the Big Cat House. Everybody wanted to see the lions being fed. Da would hook a lump of meat and carefully lift up a cage door about 10 or 12 inches. He needed great dexterity because he had to pull up the door, shove in the food and let the big cat grab it without getting attacked or having the cat come out on top of the public.

It was very dangerous. There was only a pin to stop the gate from rising, no locks. A leopard's paw would come out, and if the pin wasn't in properly, the gate would fly up and the leopard would be out. Leopards were the most cunning when feeding: they would grip the feeding fork between their wrist and paw, and try to lever the gate higher. There were no health and safety rules, either. If a lump of meat was too big to pass through the gate into Herbert the lion, Da would pull out the pin and raise the gate further. He and Herbert would be staring at each other as up to 300 people in the Lion House watched. Luckily, the lion always decided to go for the closer option, so he would grab the meat and walk away. Herbert, who was a magnificent black-maned lion, was reputed to be the biggest in the zoo world. Despite his size, Da says he would purr when given a scratch behind his ears, just like any other cat.

The visitors loved feeding time, though it meant very little for the animals aside from time to eat. As I grew up, I realised how wrong it was that some of the biggest cats in the world lived in that house with no space and no outdoor areas. These were animals designed to travel for many hours a day, able to bulk feed up to 100 pounds of meat in one sitting and not eat again for 10 days. But they were given a little bit of meat every day because it was providing entertainment.

During the 1950s and '60s, entertainment was also provided by the Chimpanzee Tea Party. This was seen by the Zoo Council as a great way of attracting paying visitors. Crowds gathered on the main lawn to watch three or four young chimps (at one point it was three chimps and Tommy the orangutan) sit around a table in high chairs and mess with bowls of fruit and mugs of milk, often stealing food from each other. Sometimes they wore aprons or human clothes. It sent the wrong message, but due to a lack of understanding of the needs of these animals at the time, chimps were so deprived of stimulation that keepers like Jack Supple and Michael Clarke could sense they looked forward to getting out. There were also opportunities to snatch a treat from another unsuspecting creature, just as there might be in the wild. Da was working on it one day as a woman in a long summer dress stood watching, licking an ice cream. Johnny the chimp ran over and put his hand under her dress; she jumped and let the ice cream fall, and Johnny grabbed the cone and ran up a tree. They loved to steal ice creams from children or to push them out of their buggies and sit in them themselves. It usually took at least three keepers to maintain control.

I was a baby in a pram when Da first introduced me and Margaret to Judy the chimpanzee. When the adults were looking away, she jumped up beside me, stole the baby bottle from under my pillow and began to suck. Da said that Judy had obviously been observing visiting mothers place their baby's bottle back under the pillow to keep it warm, so she knew exactly what to do. Chimpanzees are highly intelligent problem solvers, and they love to mimic human behaviour.

People were always knocking on the door of the keepers' room at the Roberts House to ask questions about their job and the big cats. The room had a permanent fug of smoke, and there was always a lot of banter and fun. The office is still there, with the original floor that Da washed down daily with disinfectant powder and buckets of water, but the lion cages are long gone. The wooden floors and wooden sides of those cages were scraped to pieces – it was only when the lions moved to the far side of the zoo, onto the grass for the first time, that they had trees where they could scratch their nails and work on their claws, as lions need to do.

Over the exit door in the Roberts House, you can still see the box cage that was dropped down to move a cat between cages. Sometimes, when cats were being transferred, the counter-balanced doors didn't close properly. Once a lynx slipped through a gap and out the back door. It was the same colour as supervisor Martin Reid's dog, and two women going into the zoo shop thought that's who they were greeting – until it snarled at them. Da and some others got up on their bike with nets and caught the lynx hiding in a shed on the North Circular Road.

Da was well before his time. He had a great imagination about how to improve the wellness of the animals in his care. Long before it became common practice, he would put different substrates (what is on the floor or ground of a habitat or animal house) into the lion cages to give them different scents for mental stimulation. I also began to question things from a very young age. While watching nature programmes or the *Tarzan* films, I realised I never saw animals playing with cardboard boxes, plastic barrels or car tyres like they did at the zoo. Animals in the wild were always interacting with branches, trees or soil.

The zoo also came home. We hand-reared for a variety of reasons, and we had lions, tigers, jaguars, a wolf, orangutans and gorillas in the house at different times. Sometimes I'd be fighting for my breakfast bowl with a chimpanzee. We had plenty of friends because everybody wanted to come and see what was happening with the animals. I'd come home, and there would be loads of kids waiting outside to see what the Creightons had in the house. It wasn't a Jack Russell barking at the postman – it was, more likely, beware of the leopard!

Saraswathi

In 1936, the governor of Madras in India sent a young female calf, who had been taken from the wild, to Dublin Zoo. Not long after Saraswathi arrived, she broke down the doors of her house. Despite this, she was soon regarded as gentle enough to give rides to children.

Sarah – as she was known to generations of visitors – would become one of the zoo's favourite occupants. She walked around with elephant keeper Jimmy Kenny (broadcaster Pat Kenny's father) and his father before him. Under their instructions, she gave rides, lifted her trunk and legs on command, posed for photographs and even 'played' a mouth organ. Visitors paid a penny to feed her buns. There's archival television footage of Sarah doing tricks surrounded by children who are feeding her – some of them only toddlers.

In 1941, German bombs were dropped on Dublin. While most fell on the North Wall, killing 37 people, one bomb fell in the Phoenix Park, very close to the zoo's boundary. The glass in the zoo residence was blown out, but the animal houses were unscathed. However, in the morning, keepers realised that Sarah had opened her door, gone down to the lake and returned to her house.

Da remembers buses of American tourists pulling up at the side gate of the zoo and filing in to meet Sarah. Mr Kenny would ask a woman, 'Where are you from, madam?' She might reply Texas, and Mr Kenny would say, 'Now Sarah, three cheers for the lady from Texas. Hip, hip …' and Sarah would trumpet, to the crowd's delight. Her final trick would be to take a tin plate in her trunk and hold it out for tips.

In 1950, Sarah received a much-needed companion when an elephant calf, Komali, was sent as a gift from Sri Lanka. Records say Komali was suffering from the effects of her long journey when she arrived, but she recovered enough to eat 30 shillings' worth of spring carrots for her breakfast. It was reported that, after a formal introduction to Komali, Sarah 'trumpeted with joy and henceforth adopted "Komali" as her own'. The pair stayed friends, and there are photos of Sarah washing Komali, holding a brush in her trunk.

However, nobody aside from Mr Kenny could manage Komali. He was out sick with pneumonia the day Komali charged another keeper and escaped her enclosure. Da, who was working in the camel enclosure, watched as she knocked a nun, who was with a group of local schoolchildren, to the ground. Komali's target, however, was the zoo's new arrival: a grey tractor that was puffing smoke. She lowered her head and overturned the tractor, with its driver, gardener Bill, trapped in its cab. They sent for Jimmy Kenny, who lived in Infirmary Road, and he came cycling up in his pyjamas. He roared at Komali – 'the biggest b*****d since Cromwell' was his favourite insult – and grabbed her by the ear, and the elephant meekly walked back into her stall. However, Komali grew increasingly dangerous to both keepers and the public, and in 1966 the decision was made to put her down.

Elephant rides were first given in the zoo in 1904 by the elephant Padmathi. A trainer was brought in to get her to carry a saddle, and James Kenny Snr was given a commission on takings. People still talk fondly of riding Sarah in the zoo in the 1940s and '50s, when she would carry up to eight children and her keeper in a specially designed chair, or howdah.

We know now that this should never have been allowed. There is an educational exhibit about elephants in the zoo's old Haughton House, which includes an elephant skeleton. Just one look at its upwards-shaped spine, and you will realise that they are never meant to be ridden. Any pressure on an elephant's back must cause pain and discomfort.

Sarah gave rides for over 20 years, despite severe foot problems. Grooves were put on the concrete floor of her house to ease the pressure and pain, and a special set of boots with iron soles was even made for her. She only gave rides along the softer ground beside the lake. When she tripped and fell in 1958, carrying a group of children on her back, it prompted a discussion about the way the animals were used in the zoo, but she was still giving rides a year before she was euthanised in 1962.

Sarah was part of a very important evolutionary process for elephants at Dublin Zoo. She and all the elephants that came before her are the reason why I became an advocate of modern elephant care. She is part of a past – and, unfortunately, a present – where elephants suffered for our entertainment.

I still hear people lament that they can't ride, touch or feed the elephants any more, talking about the good old days of Jimmy Kenny and Sarah. When people ask me now why they can't get close and touch the elephants, I tell them the well-being of elephants must come before making human memories.

THREE

SUNDAYS AT THE PET FARM

My education in animal care began when I was just a small kid. In addition to helping out with the hand-reared animals that came home, I was given a German shepherd dog, Dino, when I was about 10. The lion cubs used to sleep with him. There'd be times when a lion cub would latch onto my leg and wouldn't let go. I'd be going to school covered in scratches, and the teacher would ask me what happened. I'm sure they were a bit taken aback when I explained I'd been playing with a lion cub!

As animals grew older and stronger, we had to be more careful with them. There came a time when they couldn't come into the house anymore, particularly with meat eaters. It was all right when they were on the bottle and getting milk, but when they were introduced to meat, it was hard for them to distinguish between your hand and the meat. Carnivores are impulsive. There are thousands of years of evolution in their DNA. And we know now that they shouldn't have been in our world.

However, as a child, it was such a positive thing to grow up surrounded by and caring for animals. Animals are good for you, both physically and psychologically. You learn sentience, you learn to show emotion. It was particularly good for me. Back then, boys were told to be tough. You couldn't show weakness, but you could still go out and cuddle an animal. It's easier to express love for an animal than a person at that age. It's not going to snap or give out to you. It's very uncomplicated.

I always wanted to be outdoors. I was always searching for wildlife and animals, even in the city streets. A character called Lukie Nugent used to go around the houses to collect slops to feed the pigs. I waited for him every day, so I could feed and rub his horses. There were service laneways behind the houses, and people used to let their dogs out in them, so I'd be running up looking for dogs to play with. Up a couple of the lanes, there was a dairy, and a couple of fellas kept pigs and ponies. I would sit up on the wall to look at them. And it was always exciting watching the cattle run down the streets to the abattoir on Manor Street.

I found nature wherever I could. Up the road from Ivar Street was an area everyone thought of as a wasteland, but to me, it was a green oasis. I loved spending time there. I would climb over the wall, get on a roof and drop into it, looking for mice, rats and birds. I was fascinated by them. I made snares with fishing line to catch pigeons. I'd put a bit of bread inside the snare and then, when the pigeon came, I'd pull the line and catch it. I was probably hurting it, but I didn't think of that. I just wanted to look at them and all their colours.

Dublin Zoo is based in the Phoenix Park, 1,750 green acres in the heart of the city. As I got a bit older, I became increasingly drawn to the park itself, just like Da before me. Many of his ancestors had grown up around the park. His mother was born in Chapelizod Lodge because her father was a parkkeeper and gardener. He had been born in a cottage beside the Hole in the Wall pub. So the park has been at the centre of our family's lives for generations. I would go wandering with other local boys around the People's Gardens and the Wellington Testimonial or looking for deer. During the heatwaves of 1976 and 1977, if I wasn't bursting bubbles of tar on the street with lollipop sticks, I was fishing for pinkeen with nets or jumping in the lakes. Even getting to the park was an adventure. Like animals, small – and big – boys are possessive of their territory. The shortest way was through O'Devaney Gardens, a tougher area even than ours. If we went through there, we'd always end up fighting, which I used to love. I was always good with my fists, so I never wanted to go the long way around.

I was nine or ten when we moved out to Blanchardstown, and it was as big a change as moving to the jungle. There wasn't a lot to do. They made the same mistakes in Corduff as they had made in other new suburbs and housing estates: they built them with no infrastructure for children. But there were open spaces everywhere. I was in my element in the fields and around the Tolka River and valley. I spent most of my time out at the old hospital in the woods, looking for hares and rabbits. I'd hide behind a tree for hours and watch the badgers. I had a specific sett that I used to leave food for.

If I wasn't in school, I'd be up and out in the morning, and I might not come back until nine o'clock at night. My mother would be up the wall – there were no phones in those days. I'd often come back with an injured pigeon or a rabbit under my arm. In my bedroom, I'd always be rearing a couple of finches or something that had fallen out of a nest.

I used to get in awful trouble with the Travellers because I'd feed their horses or take them if they were tied up. We had a side entrance, and my ma would be shouting to my father, 'Ger, he's at it again. He has another horse out the back garden.'

I loved exploring and working around Blanchardstown, but funnily enough I never really considered it to be home. I always gravitated back to the streets of Stoneybatter, and I still went to school there until I was 15. I attended The Brunner, St Paul's School in North Brunswick Street. Everyone wanted to be my friend because if I liked you, I'd bring you up to the zoo with me.

Throughout primary and secondary school, I would bring in odds and ends – a lion's or a tiger's whisker, or an ostrich egg (I would make a little hole at each end, suck all the inside out, then swish it out with disinfectant). Da would always tell me to make sure I could answer any questions. I loved learning things so I could tell other kids about them.

The teacher would go out for a smoke or a break, and I would hold court. These were kids from the north inner city, and they got a chance to hold a lion's whisker or see the size of a big cat's first tooth. Everybody in the class would be gathered around, captivated. Schools were pretty bleak then, so it was great to bring in a bit of colour from a different world.

At weekends, the keepers would bring their kids into the zoo, like myself and my sister Margaret, and Michael Clarke's daughter Helen, who also went on to become a keeper. It was like a big family event, our own private club. We picked up litter (a very important job in a zoo) and cleaned the coins out of the wishing well. Joe Byrne, a lovely keeper that all the kids gravitated towards, sometimes let us feed the elephants.

We spent most of our time in what was called Pets' Corner, where all the small animals were kept. We'd get to bottle-feed lambs or handle rabbits, guinea pigs and dogs. It helped to build our immune systems. There was no washing your hands or using alcohol wipes then. We went from one animal to the next, eating a sandwich with one hand and feeding a lamb with the other.

Nowadays Pets' Corner has evolved into the Family Farm, where city kids get to know about farm life, how milk arrives at your door and how meat is processed. During World War II, many people donated their pets, including lots of budgies and parrots, when they couldn't feed them. By the 1970s and '80s, it was a proper menagerie of sick, injured and recovering animals – and a brilliant place to be. There was no talk of biosecurity. Everything came in: squirrels, goats, baby donkeys, abandoned fawns found in the park, doves, magpies – anything that fell out of a nest or was found injured was brought in.

Teesie Craigie and Maureen Keneally, who were in charge of Pets' Corner, are responsible for producing more keepers in Dublin Zoo than anyone else. They taught us everything they knew, as they hand-fed rats, mice and crows. We learned kindness, and to give everything a chance of life. They treated us like we were their own, even giving us dinner in the Pets' Corner kitchen, where there were always extra mouths and beaks around the table. The tame crows they had reared would come over every day at one o'clock and bang on the window to be fed. They would be let in and sit on the table beside us, taking the food off our plates. It was madness, but wonderful madness. When one crow, Joe, was a baby, he would travel with Maureen in the basket of her bike back to her flat on the North Circular Road. Once he was able to look after himself, she left him in the zoo, but on her days off he would fly up to her flat and knock on the window to wake her up.

Sometimes we had little nurseries for chimps that were being hand-reared with bottles. The chimps were popular with visitors because you could see into the nurseries, and they'd be taken out for photos. They still had the little pony and trap that my dad had driven, and when I was a bit older, I'd take visiting children on jaunts around the zoo. However, they eventually stopped the rides because there were a couple of near-misses when ponies bolted.

We didn't understand the education we were getting then. We just wanted to be around animals and, as long as you didn't mind sharing your dinner with a rat, mouse, pigeon or crow, you got along fine! There was also a great sense of freedom in being able to run around the zoo. All the keepers would look out for each other's kids – we were like a large extended family. Those weekends spawned a whole generation of zookeepers. You could pick out the children who wanted to work with animals. If you did well in Pets' Corner and showed your affinity for animals – and for hard work – you were almost guaranteed to get on the Animal Care team at some point.

In the old days, you got a zoo job because of a friend or family. There were bad keepers who had no abilities with animals, but the good keepers included second-generation zoo people like myself and my brother James, Helen and Liam Reid. We had practically been raised in the zoo. We had animals in our hearts.

It might all sound like an idyllic childhood, but it wasn't.

Like so many other families, we had our difficulties. Money was short, and I remember having very little. There were tough times, and I had to mature very quickly. From the age of about nine or ten, I always had part-time jobs or a collection of jobs, so I could give my mam a few bob.

I was always resourceful. I worked with milkmen; I washed windows; I'd dig gardens. I also had a paper run. I was delivering newspapers in Corduff in 1977 because I remember the headlines when Elvis died. One summer, I was the nipper making the tea on a building site. They were building a local school, and they asked my mam to cook for them too. They had a collection for me when I left, and it was a lot of money at the time. Jim Bolger, the horse-racing trainer, had a yard up in Clonsilla, and when I was older I went up and asked to help. I was paid a few quid a day for walking the racehorses around, feeding them and mucking out.

I had friends, but outside of the zoo I spent a lot of time on my own. I would go walking for miles with Dino because I just wanted to be out in nature with the animals. It was probably a coping mechanism when I felt insecure in my life. I still find incredible peace and tranquillity in nature – whether it's in the desert of the United Arab Emirates or County Wicklow. For me, spending a few hours around Glendalough is better than winning the Lotto. It's good for the soul, and I always come away feeling better.

Dino was my best buddy. A dog will wait at the door for you 365 days a year, and it doesn't matter whether you come in your Armani suit or rags – you're going to get the same response. He used to take my clothes off the line to lie on if I wasn't there. Animals gave me emotional stability when I needed it most. As a child, I often found that people might disappoint you, but animals never did. I always felt peace around animals. They were far less complicated than people.

I have a wonderful wife and great kids of my own now. I have always been determined to ensure that my children feel secure, both emotionally and physically. Perhaps that's also what makes me so driven, so committed, to keeping elephant and animal families together. The first thing I do when I'm assessing an animal is refer to its contribution to the family. It's all about the family: survival, learning from each other, multigenerational groups working together and, most important of all, the next generation.

FOUR

SCHOOLDAYS

After school finished for the day, I'd go up to the zoo and wait to get a lift home. It was a great excuse to spend time there. There were no health and safety restrictions, and keepers' kids would come in and out the back gate and hang around. I realise now how poor some of the conditions were, but at the time the place was heaven for me. I'd often bring a friend with me, and I'd be proud as punch walking around with them. We'd watch the lions being brought in, going up on their hind legs to get the meat off the hook Da was holding up. Their power and energy as they were ripping through the meat were amazing to witness.

The Brunner was a tough school. I was a strong fella, and I was constantly in the middle of fights, protecting other kids from the bullies. Many of them were the brightest fellas in the class, and I took them under my wing. I met lots of people later on in life who thanked me for defending them. It's always been in my nature never to allow injustices.

As a kid, I was always getting into fights on the street, too, and Da decided that I'd be better at channelling those skills properly. Avona Boxing Club in Arbour Hill was just five minutes from our house, quicker if you ran. I was a natural. I took to boxing as soon as I put on the gloves, and I won my first Dublin League championship when I was 11 years old.

If I wasn't in the yard sorting out fights at lunchtime, I used to run into town to see if anything was happening. When the *Piranha* movie came out – in 1978, I think – there were two piranhas kept in a tank outside the Carlton Cinema on O'Connell Street. I was fascinated by them and would go up every day to look at them. The staff got to know me, and I'd be telling them all about the fish, how many teeth they had, what they'd eat and where they came from. I also loved going around the pet shops. There was a great one on Capel Street and Wackers Pet Shop on Parnell Street. They were very tolerant because I would go in every day just to stare at the pigeons, and sometimes they'd let me hold them or help clean out the cages. I just wanted to be around animals.

By the time I moved up to secondary school, I was a well-known boxer, and the bullies knew to stay out of my way. This meant I didn't have to fight, and lots of the boys who were being picked on came under my wing. But it was savage in the yards. There were between 600 and 700 young fellas with nothing to do except fight for dominance or push their way around, and just one teacher keeping an eye. There was conflict around every corner.

It was a Christian Brothers school. There were some very good and kind teachers there, but others were still dishing out brutal corporal punishments, and the violence was horrific. There were punishments if you were late, if you didn't turn up with your school work or for no reason at all. We were all hit with the leather straps, and they would turn them sideways to inflict more pain. One day I stood up to a Brother who was hitting a fella in the face. I started shouting, 'You leave him alone! You've no right to be hitting him like that.' I put it up to him. I was never usually troublesome, and he knew that, which is probably why he backed down.

In class, I wasn't the most academic (more through laziness than anything else), and I already had my mind set on where I was going. I remember a teacher saying, 'When Creighton does open his mouth, only something smart comes out.' It might not sound like it, but he meant something smart or meaningful, not cheeky.

I used to love learning about other countries in geography, and I loved biology – anything to do with the outdoors or living creatures. I also liked English, listening to and reading stories. However, I was so preoccupied with wanting to be outside that I wasn't great at paying attention to anything I wasn't interested in. I certainly wasn't interested in maths. One day a teacher told me, 'The only way you'll ever get into Trinity College, Creighton, is on a messenger bike.' Years later I thought of him as I stood in front of Trinity College Dublin's zoology students giving a presentation. That was a good moment, and I still do presentations to students at Trinity and University College Dublin.

A lot of fellas in school were from very poor, deprived backgrounds. They had nothing. Many of them came to school without lunch. When I was a teenager, there was a mad trend for stealing cars, and my classmates would also be robbing radios out of cars to sell to scrap-metal places or car breakers. Fellas in the classroom would have radios in their school bags and would tell you that they got it the night before and were going to sell it at lunchtime.

There wasn't a lot to do around the inner city as a teenager. I loved keeping busy, though; work, the zoo and boxing were what kept me away from trouble. Boxing provided a great release for me, and it gave me confidence and discipline. All around me, my friends were crumbling – there was anti-social stuff like drinking and a lot of substance abuse. I never even put a cigarette to my mouth because I wanted to keep my body right for training.

Once again, the Phoenix Park played a central role. It was the best ready-made gym in uncontaminated fresh air, and it gave us an edge over most other boxing clubs. We'd run up to the Castleknock Gate and back every night. On Sunday mornings we'd leave the club, go up through The People's Gardens, around the zoo and back around the park. Later, when I was on the Irish squad, our training was across the road in the Garda depot. I would walk out the back gate of the zoo, go across to the depot for training and come back again.

Another great place to train was Magazine Hill, near the Magazine Fort. It has a whole series of challenging topography that goes up on a slant, and it's a great place for building the leg power essential in boxing to help absorb the shots. Or we used to go over towards the Furry Glen area, where there's a beautiful lake and a sequence of steps that we would run up and down.

In addition to training at the Phoenix Park, I'd go for a run every morning with Dino, or I'd cycle out to the Strawberry Beds to swim against the Liffey weirs. Swimming against the water as it pushed me down the river was great for building body strength. I'd do that summer and winter. Now I realise how dangerous it was: at 16 I'd be down there on my own, and anything could have happened. I was fearless at that age.

When I look back on it, I know boxing saved me because it gave me a purpose. Drugs got a hold in Dublin because there was no mental stimulation for young people and no infrastructure. Up around O'Devaney Gardens, even the playgrounds were vandalised. Fellas were going out of school and up to Grangegorman smoking hash, and there were even a couple of guys taking heroin.

It was awful to see the physical deterioration of people my own age. I would chat with everyone and could see the habit take hold and destroy them. I was offered drugs regularly, but it was never something that appealed to me. I always wanted to go to work, help at the zoo or box.

A lot of fellas I grew up or went to school with ended up in prison for one reason or another. Once, as a keeper, I went into Mountjoy Prison to do some talks. Before I went in, the prison officers warned me: 'Listen, Gerry ... they can be a bit messy, and they can try and test you. There'll be two or three guards with you, and they'll interject if it's needed.' But I walked into the room, and loads of the inmates recognised me and were high-fiving me. I'd say I knew about 90 per cent of the lads there – a lot of them since they were kids. So the guards went for a cup of tea and left me with them. I was only supposed to be in there for an hour, but four hours later, they were showing me their cells and their artwork. I went back on three or four occasions after that, bringing down a load of reptiles and snakes and educating them in whatever way I could. It was good to be in the 'Joy for the right reasons. Those lads were no different to me – they just didn't get the opportunities I had.

I have been lucky in my life to meet good people who have taken an interest in me and helped me to build a sense of responsibility. Eileen O'Connell was one of those people. She had a little record shop and video arcade in Blanchardstown called Sounds Cool. Eileen took a shine to me when I was about 12 or 13, and she treated me like her younger brother. She encouraged me and gave me a little weekend job. She would even leave me to look after the shop. All the DJs would come in from radio stations, and I'd go to EMI and the record distributors when she was buying stock for the shop.

Eileen had a huge influence on me. Even now, over 40 years later, she still sends me a Christmas card. Her father Jimmy had a farm, Rosemount, which was only about a mile from where I lived, and I would go there for dinner. He took me to cattle marts around the country and gave me a job working on the farm. At only 14, I was earning 20 quid a week and bringing it home.

I was always thinking beyond school. I did my Inter Cert, but the day they blew the whistle and it was over, I ran straight to the zoo and vowed never to leave it again.

I was 15 when I left school, so I had to do a pre-work course for a year at Lucan College while waiting to get a full-time zoo job. When a glazier apprenticeship came up and my mam wanted me to go for it, I told her I couldn't risk cutting my hand as a boxer – but really, I just wanted to work in the zoo, no matter how hard or how badly paid the job was.

My career as a pugilist had almost ended in my early teens. I had heard a rumour that they were testing on animals in the Department of Agriculture's research facilities at Abbotstown, so I headed over there to 'save' a horse. It was a racehorse, and I planned his escape carefully. I fed him to get him settled, then cut through the fence and jumped on his back. The next few seconds were like an eternity. A bucking bronco would have been outclassed. The laws of gravity naturally prevailed. I was thrown off, knocked unconscious – the only knockout I ever experienced – and smashed my left hand. I made my way to the hospital alone (my parents didn't know what happened until later) and required surgery. I was told I would never box again. However, Da got me to squeeze a tennis ball to rebuild the muscle, and I was soon back in the ring. I went on to win lots of titles.

I was also put temporarily out of action by a head injury when I got hit by a car while cycling to Lucan. I was blessed. There were no bicycle helmets in those days. I got 16 stitches to the head, but it didn't keep me out of the ring for long.

I boxed throughout my teens and into my twenties. I was a Dublin and Leinster champion, the national youth middleweight champion twice and the under-18 Irish middleweight champion. I also got to represent Ireland internationally.

But then I came to a crossroads. I was working at the zoo nearly every weekend, and it was harder to get time off. It was never a job where you could say, 'I'm going to finish at half-four, go home for a cup of tea, and then I'll go boxing.' It was: 'Gerry, you're needed here' or 'Gerry, you're needed there.' In the summer, I could be working till 7 p.m., then I had to go down to train. I wasn't eating properly, and I was tired. Or there could be a sick or newborn animal, so training went out the window for that night. I just couldn't give the commitment level that was needed to be a successful full-time boxer. It wasn't that I didn't want to: I just couldn't give it.

And I needed money. I had to take holidays or time off work to train. I didn't get paid for my time, and that was always a struggle. As I got a bit older, I also wanted to go out and socialise. I would chance going out and having a few pints, even when I was training. Working at the zoo and being an Irish champion made me a great hit with the girls – I'd either have the boxing gloves or maybe a lion cub in the back of the car.

So I had to make a choice: did I want to be a boxer or did I want to be in the zoo? I was realising that I had inherited a gift with animals from my father. I wanted to be around them, and I wanted to change things for the better for them. So I chose my career – a deliberate choice, rather than simply drifting into the zoo because my old man had been there.

Judy and Kirsty – 1

Judy and Kirsty were the last elephants to be kept in Dublin Zoo in free contact (meaning we would go in and share the same space with them, carrying a hook, or ankus). It was their history that inspired my desire to improve the lives of elephants in human care. Both had been born in the wild in Southeast Asia: Judy in Thailand around 1957, and Kirsty 10 years later. They first came to zoos at the age of about four.

Judy arrived at Chester Zoo in 1961, and her only calf, Jubilee, was the first elephant to be born at the zoo in 1977. Jubilee's father Nobby never even got to meet him: he was shot after climbing across the enclosure's dry moat and breaking through the zoo's fences out onto a busy road. He was believed to be in musth at the time of his escape – this is when a large rise in reproductive hormones can trigger aggressive behaviour in bull elephants.

Jubilee was later transferred to Belfast Zoo, where he died in 2003 during surgery on a broken tusk. His skull can be seen in the education centre at Dublin Zoo.

Kirsty lived at Calderpark Zoo in Glasgow, Scotland, from 1972 to 1987, before moving to Chester Zoo, where she met Judy. They lived together until 1991 when Judy moved to Dublin Zoo, but Kirsty followed three years later, and they were together again. Kirsty had a lovely, placid temperament, while Judy was the leader; she could be very bossy with Kirsty, particularly if there was food around.

The Elephant House in the 1990s suppressed every natural emotion of the elephant. It was confined, it was cold, it was dull, dark and horrible, and it was totally unsatisfactory for elephant wellness. It was so small that, when we came up in the morning, Kirsty would be all stained down one side from trying to lie down on the ground. She was housed in such a small area that, when she defecated or urinated, it covered the floor. Judy was slightly arthritic and never lay down, except one time, when we put some sand hills outside, and the hill could support her getting back up.

During the 1980s and '90s, children's parties were given access even to the elephant and hippo houses to generate revenue. It's a miracle that we never had an accident, but that was testimony to the keepers that worked there, particularly Joe Byrne, who had a fabulous relationship with the animals and kept them so gentle.

The party would be brought to the pool, where the birthday kid would be allowed to throw food into the hippos' open mouths. Luckily, Judy and Kirsty were very calm elephants, but there might be 20 or 30 kids and their parents in their home, touching them and trunk-feeding them. It was not good for them, and potentially extremely dangerous. I shiver at the thought now, but even then I was never comfortable with it. However, it was part of the job.

For me, the biggest worry was that when I went in there in the morning, I carried a hook. It always felt so wrong to have it in my hand, and I never used it. However, it used to kill me that these animals associated my presence with pain. We were lucky that Joe had developed a relationship with Judy and Kirsty that allowed us to share their space without dominance. However, while we didn't need to use force on Judy and Kirsty, I knew this was how they had been trained – with pain as punishment – and it made me feel very uneasy. It was all wrong.

Judy and Kirsty were so smart in finding ways to get what they wanted. Before health and safety rules insisted on keepers working in pairs when sharing spaces, I could be on my own with the elephants at the weekends. I would be outside with the wheelbarrow collecting elephant poo (one of the main jobs when working with elephants, since they produce over a tonne a week!), and, if it was in any way cold, Judy would come behind and delicately push me. Despite her power (and weighing four tonnes), she would manage to be extremely gentle, moving my body ever so slightly. She would push me left, right, left, right, forward… until she had moved me all the way to the gate, to show she wanted me to open it and let her indoors. She would want to go in because it was cold, she knew the feed was inside and there wasn't a lot happening outside (at the time it was just a bare piece of land with a pool). I might give her some hay or nuts, but she'd turn me around with her trunk, then ever so gently push me the way she wanted to go.

When the African Savanna opened, and the old giraffe and hippo houses were demolished, Judy and Kirsty were given more space. However, as she aged, Judy became more arthritic, more resistant to doing what we wanted (including allowing us to give her essential foot care) and more temperamental with Kirsty. At the time, keepers were being killed every year while sharing the same space with elephants (and often older, arthritic elephants). One day Judy went for Kristy, and I was a hair's breadth away from being crushed. I had to move out of her way very fast, or I would have been dead. She wasn't after me – I was just in the way.

We began working with world-renowned elephant expert Alan Roocroft to make improvements. Large tree-root balls were brought in for them to play and interact with; we got some browse feeders to hang in their house; and we introduced some protected contact training. But we realised these were only plasters on gaping wounds. The elephants' needs still weren't being met. The danger of being confined together in their house also became apparent during Halloween 2004, when they became stressed by the sound of fireworks exploding. When their house was opened in the morning, both were covered in scratches. Kirsty had a bite wound on her tail, and Judy's tushes (the teeth next to the tusk) had been snapped off.

Leo Oosterweghel, who had just come in as zoo director, looked at the facilities and decided we needed to start over – we couldn't continue to manage elephants the way we had been. When an opportunity came up for Judy and Kirsty to join an older female in Neunkircher Zoo in Germany, we went out to have a look and saw that it would be a much better home for them. It was a new start for them – and for us.

FIVE

A JOB AT THE ZOO

When I was 15, I finally started being paid to work in the zoo. At first, I worked part-time with the ponies in Pets' Corner, mucking out the pigs and feeding the chickens. I also cleaned the grounds and worked with the gardeners.

At 16, director Peter Wilson offered me a trainee keeper position. He could see that I was a hard worker and a natural animal person. I was working alongside my dad, mainly with the big cats, apes and elephants, but I also moved around to provide cover in other areas when staff were out sick or on holidays. You had to be committed. I worked weekends when all my friends were going out and doing things. It wasn't unusual to go six or seven months without a day off – Da was the same – and it continued like that for years.

Even when I was out with my mates, I always had it in the back of my mind that I would have to leave early because I had to be at work in the morning. However, it was never a burden. I realised from a young age that the animals were depending on us. In those days, they had nothing in their lives but the keepers. We were the focus of their day. That responsibility was not lost on me, and that was a lot to do with how my dad taught me.

My very first day as a trainee keeper was spent at the Elephant House. I was very familiar with the house, but on that day I looked around and realised how miserable the elephants must be. These were African elephants at the time, and neither the facilities nor the keepers were equipped to care for them properly. The zoo was very poor, and the elephants would just be given some hay at night and some fruit.

Because I was physically fit from boxing, all the keepers used to ask for me to go to their sections for difficult or heavy work, like a deep clean of the Elephant House or the pools. I remember telling to my dad I was getting pulled all over the place, and he told me, 'It's a good sign when people are asking for you. It's when they are not asking that you should worry.'

Because I was considered reliable and versatile, I was able to step in anywhere – I could be with the sea lions or in the Monkey House. I was a floating keeper for a number of years. Now, when people say they want to be an elephant carer or a lion keeper, I always advise them to gain whatever experience they can with a broad range of species. A good animal carer works with all the species at some stage. You need to get an understanding of every animal and learn different skills. A good carer understands the ecological role of all animals: how the fly lives, how the butterfly eats, what the chimpanzee needs … all the way up to the elephant.

Being this flexible early on in my career was an invaluable learning tool, and I still surprise myself with the information I can recall about different animals and different species. It was the making of me as a keeper: one day I would be carrying a 20kg piece of meat or the leg of a horse to feed the lions, and the next I would be chopping minuscule pieces of fruit and veg for the monkeys. I loved that variety.

It's always interesting to see the different personalities of zookeepers who work with particular animals. Ape keepers often have dramatic personalities, expressive and full of talk. The big cat keepers have a swagger in their step. Reptile keepers, meanwhile, tend to be calm, relaxed and solitary – like snakes – but they are also very fast, because they may have to react quickly.

Dublin Zoo is the fourth-oldest zoo in the world, after London, Paris and Vienna. When I started, it housed animals so they could be gawped at by visitors. Often our residents' only stimulation was feeding time. As we evolved in our thinking, and our recognition of animal needs, we started to give them ledges and make improvements. However, we were still left with the inheritance from previous generations, where the key guiding forces for design were containment, confinement and cleanliness. Everything was bleached and cold. The conditions were dreadful. It was about keeping people safe from the animals and stopping the animals from harming themselves. They never considered the animals' biology and physiology, or any of the key components for animal well-being, welfare and mental stimulation.

When I started, the set-up of the zoo was dictated by what was available through animal dealers and what people had seen in the *Tarzan* movies. Most zoos had similar collections. The

Chipperfields, the circus people, were the main animal dealers and trainers, supplying parks around Europe and the UK. Someone would ask for ten elephants, and the Chipperfields would send a team to Africa to get them. That's how it was done. It was so wrong.

At that time, animals were allowed to breed consistently. The lions, bears, kangaroos and wallabies would produce babies every year. The Chipperfields would come in when the season was over, in autumn, to buy and move animals. I would ask where the animals were going, and I was usually fobbed off and told they were going to some safari park. I don't know where they really ended up.

The way we treated animals was not welfare-based. Young lions would be given a mild sedation, and I'd be pushing them by the tail or grabbing them by the scruff of the neck, while still awake, to get them into boxes. It wasn't good for the animal. Keepers that fed and looked after the animals suddenly came in, threw a net over them and forced them into a box. I often thought about how confusing it must have been for these animals.

In the past, being a zookeeper was a very dangerous profession, though in Dublin's history, only two keepers were killed: one in 1903 by an elephant and the other by an Irish red deer stag. The former, James McNally, was killed by Sita the elephant while he was applying ointment to her foot, which had an overgrown nail. An order to destroy Sita was issued. When attempts to kill her with a poisoned apple failed, the head of the Royal Irish Constabulary offered to execute her. Afterwards, the foot used to crush her keeper's head was cut off and presented to the RIC.

When I started, keepers needed a very different type of skill set than today. The aim was to come into work every morning and get out alive every evening, without getting bitten or scratched. The work certainly honed your reflexes.

All of my boxing strength was needed for moving the kangaroos. At that time, we had eastern grey kangaroos, which are massive animals. We would rotate them between two grassy areas on both sides of the lions on the far side of the zoo. However, we had to physically handle them to get them into crates to be moved. I would get a bucket of fruit and nuts, and they would follow us up to the top of this steep hill where there was a 40-foot container that was used as their house. It had a feeding trough, so they were used to the routine of following us in to be fed. Then we would have to close over the steel door of the container – the only light was coming in from a small vent – and pin them with a yard brush or catch them. They would go back on their tail for balance and kick forward, so we had to watch out. We would grab them by the tail – which was as thick as a man's arm, just pure muscle – and they'd be pulling us around like rag dolls. Then we'd open the door and manage them down the end of the hill, while they were still pulling and trying to turn and grab us. They had formidable claws on both hands and feet that could do a huge amount of damage. Having to do this a few times a year got to be really challenging!

Being in a dangerous situation was very common for keepers. There were often only inches of mesh between us and an animal, and it mightn't be fit for purpose, so the risks were huge. But my instincts and reflexes sharpened very quickly, and I learned from those around me. A good keeper could predict situations long before they happened. Sometimes I'd think Da was being a pain in the arse when he repeated instructions, and I might mutter something under my breath. But most of the time I'd listen to him because he had a sixth sense around the big cats. He had a great foresight for danger and taught me to think things through and evaluate everything before making a decision. And these were important life-saving lessons, like always having two doors between myself and a big cat, or when dealing with gorillas and orangutans, having two doors between us that they couldn't lift. These primates have the ability to lift doors – and that did happen, on more than one occasion.

Before safety locks and devices became compulsory, big cats could use their noses like suction cups to lift doors and get in on keepers – there were quite a few near misses over the years. I became really alert and spatially aware when I was working. My father drilled into me to move to the left or the furthest point away from a dangerous animal, and that became second nature. When I spent time in the National Zoo in Washington, DC, they told me I was 'one hell of a keeper' – which was really a tribute to the time and the investment that my father put into me, as well as some innate inherited skills.

I loved the animals that triggered my adrenalin, like the big cats, and the ones I could read, like the chimps, orangutans and gorillas. I always had to think quickly, because they would try to grab me. I remember working with Michael 'Mickey' Clarke, a really knowledgeable and laid-back keeper, in the Monkey House. His brief included the chimps, including former Tea Party chimps Wendy, Betty and Judy. The old chimps' pit was like a dungeon, with awful cell-like areas at the back where they were kept at night. I hated it. It was more janitor work than zookeeping. I'd go in there in the morning, clean it, wash it out, put the food out and let them in. I had to give them a warm drink, warm Ribena and milk, and Mickey would say, 'Watch it, son, watch your hands.' If you weren't feeding them fast enough, they would spit or try to grab you through the bars. They were just so frustrated in their environment.

After a few days of training, Mickey said to me, 'There you are, son: there's the bucket of food, go on down and bring them in. They will either like you or they won't.' I went down, opened the door, pulled the bar and closed the door behind me. There was a chain you pulled down. Wendy, one of the more dominant chimps, came flying in and went for the door. I had it locked, but Wendy went straight to the lock and grabbed it. She was testing the lock to see if it was open. She realised the experienced keeper wasn't there, that I was down there on my own. If the door hadn't been locked, she would have tried to come out on top of me.

That day, I began to realise the level of intelligence I was dealing with: these were streetwise chimpanzees. They had all day to figure things out, as they had nothing else to do. The only way to get them to come in or out was with food. That was the way we had to control them. There was no training. I remember the older keepers shouting and roaring at them if they didn't want to go out on a cold morning, running at them and making noises to try and frighten them outside. I really regret that I didn't try and do more for them at the time.

Sometimes we didn't see danger, even when it was staring us in the face, and it's only on reflection that I wonder about the situations we put ourselves in before the health and safety of both keepers and animals became a priority. Hippos are among the most dangerous animals. They kill more people in Africa than any other animal, including lions and crocodiles. People go to the river, maybe to wash clothes or cross, and they might be attacked by hippos aggressively protecting their territory. The old Hippo House was really insufficient. There were two stalls, a small pool and a bigger pool outside that was barely heated. It certainly wasn't adequate for hippos, who are pretty susceptible to the cold. Because hippos use their tail almost like the propeller of a helicopter as they're defecating, poo goes everywhere. As a result, the enclosure would have to be washed every day, and the pool cleaned.

The senior elephant and hippo keeper at the time was Joe Byrne, who was around the same age as my father. He was a jolly man, with a calm disposition and a lovely nature. He was very caring towards his animals and built a super relationship with the hippos and elephants, who were almost like pets with him. Joe was also a man of routine. Everything happened at the same time every day, and everything was put back in the same place. Having such a good routine made the animals quiet and predictable, and our hippos Henry and Linda were used to the same daily routine. Nowadays we like to surprise animals with enrichment or changing feed patterns, but Joe's approach worked for the facilities we had at the time.

At feeding time, I would go in, throw some hay and food inside the house and call the hippos out of the pool (I had already opened the drain to allow the water out, so they really had no choice). They would walk up a short ramp into the stalls. Then, when they started munching on their food, I would climb through some bars and get into the pool to clean it. So I was down the end of the ramp in the pool, with the hippos – four tonnes between the two of them – looking down at me from three or four metres away. There was no way out. If they had charged at me, I would have been killed. It is a tribute to Joe's care in keeping those hippos so calm and tame that this never happened. I still had to rush to get the pool cleaned before they finished their lunch or breakfast, so I could get out in time and refill it. Thankfully, health and safety has changed so much. A keeper would never be in that situation today.

I've only ever had a few scratches and bites, mainly from hand-rearing big cats who had very sharp claws. However, I had a few close calls. Because I was the son of Gerry Creighton Snr, the big cat keeper, I was entrusted with dangerous animals from a very young age. One night, another keeper hadn't locked an inside door properly, and a puma got out into the keeper area. When I went to open the door the next morning, I registered a shadow that looked a bit unusual. Something in my brain prompted me to open it very slowly, and just as soon as there was a gap, a paw came out and caught me right on top of the knuckle. I was stuck trying to hold the door closed with my right hand while my left hand had a claw stuck in it, but I held on. I knew if the puma had got past me, it would have been straight into the zoo. I eventually got the door closed and called for Da.

My safety always depended on other keepers, just as theirs depended on me. In the early days, I had another close run-in with an American black bear. There were massive steel gates around the bears' den that you'd have to pull down using a lever. I was going in there one morning to clean up, and the keeper assured me they were outside. I slid back the gate and was about to step in when a bear came out of his night den. He was about a metre away and started to lunge at me. The next thing I remember I was outside the area with the door closed. Luckily, I had enough composure to do that. Being good in emergency situations like that comes from a combination of being in the zoo from a young age, the confidence instilled in me by my father's teaching and boxing – because if you didn't react quickly, you got your ass whipped!

Another way I improved my keeper skills was through helping out at post-mortems. Me and another fella from the maintenance team, Jimmy Quinn, used to be called to help out at post-mortems when the animals were large. I remember being in the old Elephant House, sitting inside an elephant carcass sawing through its ribs. They were so big we had to cut them up to get them outside. Samples would be taken of bone, blood, organs and tissues, anything that would help identify what had caused the death.

I was always fascinated to see an animal's structure, how it looked under its hide or fur. My dad was very skilled at skinning. He had done it for years because the zoo used to kill their own horses and ponies to feed the lions. One time, when I was very young, he was skinning a dead jaguar for education purposes, and I asked if I could come along. These powerful South American cats are the third-largest in the cat family, after lions and tigers, and they are amazing swimmers and climbers. I was so impressed by the definition of its muscles. I suddenly understood how it could project itself up a tree or jump down on prey, and the strength of its bite force. And this was a jaguar in human care. I couldn't even imagine how much more defined and powerful those muscles would have been in a wild cat.

Post-mortems were always more difficult when they were on animals I had looked after, but I was still happy to help out, because of what it would teach me. I realised that the secret to understanding an elephant is knowing what's inside it: seeing the massive blocks of teeth that grind down the food, looking at its digestive system, examining the pelvic bone that the calf needs to get over to be born.

I still encourage keepers to go along to post-mortems, to see what animals look like and how they might change when they are old, sick or emaciated for some reason. This can help us to understand their needs throughout their lives.

SIX

THE AWFUL 1980S

Dublin Zoo in the 1980s was completely different to what it is now. Back then, the lake had only a couple of small islands, including one for the acrobatic whooping gibbons. When you entered the zoo, you would first see the llamas, followed by the rhinos in their barren area with two cold stalls, and then the blackbucks, an Indian antelope species. The Orangutan House was barren concrete with a couple of bent bars that formed a sort of climbing frame. The orangutans would sit there all day, devoid of any stimulation. There was a flower bed in the viewing area of the Orangutan House, and occasionally a bold kid would pull out a plant and throw it into the enclosure. One day, I watched Adam the orang spend the day pulling a plant apart, dissecting it, finding insects and messing with the soil and roots. Da and another keeper were giving out because they had to go in and clean it up, but I was fascinated and wondered what it would be like if Adam had plants to play with all the time.

Sibu and Leonie were two orangutans who came to the zoo in the mid-1980s from Rotterdam, which had a nursery for orphaned orangs. They had been hand-reared and were extremely friendly. I was a trainee keeper, and I'd take them off around the zoo on my back or by the hand, to get them out and break up the boredom. They'd sit with us while we had a cup of tea in the back of their house. Sibu and I played together – two redheads wrestling. Leonie could be very affectionate, and Sibu would get jealous – I think, with my red hair, he thought I was a rival! He went away for seven years on a breeding loan to the Netherlands, but he's back now, and he is one of the zoo's great characters. Every time he sees me coming, he runs over towards me, but he still puffs himself up, and he often doesn't like it when any of the females look at me. If I go into their house, Sibu will come over to me and put his hand up to mine against the glass, to greet me and let me know that he sees me. Leonie will make faces to catch my attention. They have incredible empathy and intelligence.

Animals were constantly used as props or hired out for events to generate money. In the 1980s, Da was sent out to a launch with Leonie; he had to get up on the back of a Jeep with Miss Dún Laoghaire and drive around. Just before he got on, Leonie weed all over him, socks and jocks. Everything was wringing wet (and orangutan urine is particularly pungent), so Da had to apologise. 'Sorry, I don't always smell like this ... It's not my aftershave.'

He decided to try and train Leonie to urinate before he took her out using positive reinforcement, long before it became common practice. It was a long process, involving moving her from one cage to another, and rewarding her when she urinated. But it worked – sort of. Later, he and Leonie were at an event for children with special needs. They had been driven by John O'Connor (the zoo's giraffe keeper). Da was surprised when Leonie jumped off him and onto John. After peeing all over John, she jumped straight back into Da's arms!

There was nothing to interact with in the Polar Bear House either, and these private creatures were stared at all day. Even back then, when the public wasn't as educated about animal welfare issues as they are now, visitors were uneasy about the bears, particularly the polar bears, who could be seen rocking back and forth and from side to side. There was also a lot of internal scrutiny about their well-being and conditions. Concerns about Dublin's polar bears were even raised in the European Parliament. Ootec and Spunky had been captured as young bears from a dump in Manitoba, Canada, and sent to Dublin in 1980. They were considered pests in Manitoba, but they became a tourist attraction.

The pool of green, filthy water for the polar bears came from the zoo lake. There was no proper sewer ring, and the waste from the animal houses used to flow into the lake. I remember trying to do things to stimulate them, but there was nothing I could do. Sometimes they would have to be starved for days to get them to venture into the back area, it was so barren. Then we'd go out and clean the bones and debris out of the pit. We sprayed it down with a small mobile fire engine, but we were just washing the shit straight back into their water – that's how bad the infrastructure was then. I remember being in there cleaning with my father, and the bears would be banging on the doors, just banging and banging, and it was terrifying. We couldn't wait to get out of there because it wasn't safe.

What those poor bears had was a type of psychosis called stereotypical behaviour, and it's a coping mechanism for boredom and lack of stimulation. This rocking from side to side is often observed in circus elephants and some zoo elephants with inappropriate habitats and care. Attempts were made to improve their lives, but the bears' habitat did nothing to allow for species-appropriate behaviour. Ootec and Spunky were a male and a female, and they shouldn't even have been together. Polar bears usually live solitary lives. In the wild, males and females come together for one day, they mate and say good luck to each other. You might see groups around dumps or a good food resource, but they're very much a solitary nomadic animal, not suited to being kept together in a confined area like the one in Dublin Zoo. In particular, it was very stressful for Spunky, the female. It also had tragic implications – on two occasions, newborn cubs were killed by their father only moments after being born. It was awful.

The lion enclosure in the 1980s was a big step up from the old Lion House. However, the lions' routine was still far too structured and boring. They would get out in the morning and sleep for most of the day. Then, at about three or four o'clock, they'd start moving up along the fence line waiting for the keeper, because they knew they were going to be brought back in for the night to be locked in and fed. There were holding cages at the back where they came in every night, and there were gaps in the bars. The lions would be up on their two back legs waiting for keepers to come past, ready to try to pull them in against the bars. We could never let our defences down.

The old sea lion pool had only become filtered in the late 1970s, but now it was based on a human swimming pool. It looked clean for the people, but the sea lions had terrible eye issues because of the high levels of chlorine. The old Monkey House had cold tiled floors, sterile square cages and a few steel bars that were made to look like branches. The outside areas were concrete and had more bars. Older people might remember when even the large apes were kept there, including gorillas and orangutans. In those days they had a TV monitor in the house because it was seen as great enrichment to let the apes watch *Planet of the Apes* or the news. It probably was a form of stimulation for them, but it wasn't what they actually needed.

We weren't allowed to put in anything that might dirty the cages, only a little straw or bedding. The reason was to make sure the monkeys and apes were always on show. If there was too much bedding, they would hide underneath it and sleep. Given the chance, they would submerge themselves completely. Even though the Chimpanzee Tea Parties had ended, our chimps had been so humanised that they still begged for food. They'd have their hands out trying to gather nuts or whatever anyone threw at them. This humanisation also had other implications, because it meant they were not very successful in raising their young.

Up the hill was the old elephant facility. Originally it was just a rounded pit with spikes to keep them back from the public. It was barren, and the elephants were brought back into the old house every day, with its bitumen floor and no natural light. There was no stimulation for these complex, highly sensitive creatures.

When I was younger, there were quite a few elephants coming to Dublin, many of them from dealers – and a lot of them didn't last long. This was before international regulations and structured international breeding programmes, when you could just buy an elephant. Naturally, everybody wanted an elephant in their zoo. I remember Judy and two youngsters, Coffee and Walnut, that were sponsored by HB, and there was Mimi, who had been gifted to President Patrick Hillery when he was on a visit to Tanzania in 1979. I think there was a row about who would pay the shipping costs (about £4,000 at the time), but she arrived at Dublin Zoo in May 1980. She was moved to Southampton Zoo in 1982.

I always had it in the back of my mind that something was not right. I knew we had to do better. In the mornings, the animals would pace back and forth, back and forth. The giraffes, the elephants and the big cats would just pace as they waited to get out of the confined space they had been in all night. Once they got outside to the air, the keepers would go in to clean or wash down tiles and surfaces and make all the inside areas and the houses spotless again. The constant smell of bleach must have been so disgusting for those animals. It was such an alien smell for them, and they were forced to live with it.

Some of the keepers were almost as institutionalised as the elephants. They'd come in in the morning, let the animals out and then, when all the morning cleaning work was done, the old-style keepers would get into their navy uniforms (which were almost like prison warden uniforms, with silver buttons, shirts and ties) and peaked hats, and they'd stand and engage with the visitors in their area. At around four o'clock, the animals might start to pace again, watching for a glimpse of that keeper, which meant food was coming. It was the same thing the next day and the next day. The routine never changed for keepers or animals.

One of the things that sticks in my mind from my time as a young keeper was how important I was in their life. It was a wonderful feeling to see the lions' heads come up as I appeared in the distance. It felt great to be the centre of attention on a Sunday when the zoo was packed. As a 16- or 17-year-old in my zoo uniform, I'd walk up thinking, 'This feels great. The lions are looking at me, they're waiting for me to open the lock.' When they ran along the fence line beside me, people would marvel, thinking that this was some unique wonderful relationship between me and the big cats (or monkeys, or elephants), who were following my every footstep, their eyes totally fixated on me.

I'd walk with pride as the animals followed. However, it wasn't a great relationship. The animal was hungry, or bored, and I was the only focus it had that day. I was just a food resource. Looking back now, I know how awful that was. The next few minutes were the whole highlight of that animal's entire day in the zoo. I was about to bring them into a dark little area, feed them and then turn the key and walk away. It's terrible to think that this was as good as it got for them. There was no such thing as enrichment, no appropriately designed habitats where their biological, physiological and psychological needs were being met.

There was a lot of sickness among the animals because the zoo itself was sick. People were becoming more critical: they were seeing things happening in other zoos, and watching how animals behaved and lived in nature programmes, yet their zoo had barely changed in a hundred years.

I remember at least six elephants dying. There were two young African elephants brought in when I was about 15 or 16, and every day one of them would collapse. A whole team of

18–20 keepers would have to go to try and get the elephant up off the ground. It was heartbreaking. These were just some of the mistakes of the past. They are not a reflection on anybody – we just weren't equipped to give the animals what they needed.

Morale was very low, and there were lots of internal industrial relations problems, too. There were some really good keepers. And most of them were dedicated – they just needed to modernise. Despite the conditions, many of us cared deeply for the animals and always had their welfare at heart. Ríona was an orangutan we had hand-reared. Now older and back with the group, she had to have surgery, and she was opened all the way down her belly line. With apes, that kind of invasive surgery is virtually impossible because they keep pulling at the stitches. But me, Da and another keeper, Liam Reid, slept in the house with her, in the straw bed, every night for a week until the wound healed. We'd be in our sleeping bags, and she'd be between the two of us. We'd each hold one of her hands to stop her from pulling up her bandage and going at the stitches. We'd sing her to sleep and take turns nodding off. Ríona was such a sweet animal. It was great to be able to save her life like that, and we were happy to spend night after night with her holding her hands.

When Peter Wilson took over from Terry Murphy as director in 1984, Dublin Zoo was a depressing place that needed help. Peter, a veterinary surgeon who had been a member of the Royal Zoological Society Council since the 1970s, had always had time for me when he met me as a child around the zoo, and he and his wife Jane continued to be extremely kind and encouraging to me as a young keeper. Peter was a very smart man. After he made some improvements, he brought in two giant pandas in 1986 and got sponsorship for the exhibit. At that time, China had lots of pandas out on hire. It was their way of getting foreign currency, and for zoos, it was an opportunity to get lots of visitors. Ming Ming and Ping Ping were only on loan for 100 days, and they were placed in the old gorilla area. The panda was the World Wildlife Fund (WWF) emblem, the ultimate conservation symbol, so it created a huge amount of interest because people were never going to get the chance to see pandas again.

I was only 18 years old, but I was selected, along with Da and Liam, as part of the team that would take care of the giant pandas alongside our Chinese colleagues. I used to think that pandas only ate bamboo, but the Chinese team also made up a concoction of boiled chicken for them, like a Chinese soup with carrots. Their bamboo was grown especially at the John F Kennedy Arboretum in Wexford.

Looking after the pandas was quite boring. Ming Ming and Ping Ping didn't do a lot. They were asleep a lot of the time. However, that wasn't the point. The point was that this iconic symbol of conservation had come all the way from China. People didn't mind that they weren't awake, they were just happy to get a glimpse of them, and I was delighted to be involved. Before they left, the Chinese team even gave me a gorgeous gold medal with a panda on it as a present.

There has been great success in growing the panda population since then. It was very difficult because the female comes into heat for maybe one day, and then she has to be lucky enough to meet a male on that day, so that's why the numbers dwindled. What many people don't know is that a lot of the information that was learned from pandas about reproduction, ovulation, sperm collection, egg harvesting and embryo transfer was used later for humans.

Peter Wilson's idea was to show how successful the zoo could be if it had attractions. Nowadays, bringing pandas on tour like that would be controversial, because you're moving animals around the world, but at the time it was seen as a huge success. And it brought people back into the zoo. We followed that visit in 1987 with the very rare and beautiful Chinese golden monkeys, then we had some koalas on loan from San Diego Zoo. They came over with American keeper Vickie Kuder and were kept in the old Roberts House.

Peter and the board were trying to get in animals that would attract visitors and create revenue with tickets and merchandise sales to get us into a healthier financial situation. At that point, we were basically living from week to week. With this kind of revenue stream, we could see that there was potential for a future. These visiting animals also brought in visitors at quiet times. Back in the 1980s, the popularity of the zoo was seasonal: St Patrick's Day kicked off the season and September ended it. For the rest of the year, we were just keeping the place warm.

Bertie Ahern was a local TD, and he was always very involved in the zoo as the Minister for Finance and then Taoiseach. Because of him, FÁS (or AnCO, as it was called at the time) – the national training scheme – was brought in to work on building projects. The zoo provided the materials, and FÁS provided the labour. The old aviary was converted into a Bat House, and a lot of the houses and old areas were modernised. This was kind of the start of the change. Work began on a new Monkey House, but that wasn't completed because, by 1990, the Zoo Council could no longer afford the building materials.

But all those visits and new projects didn't stop visitors, media and the staff from questioning conditions and voicing animal-welfare concerns. People were becoming more educated about the natural world and the needs animals have. I was proud of what I did, but in the 1980s and into the '90s, keepers got in the habit of taking off their uniforms before going home, because they were afraid of the reaction they would get in public. It was a bleak time in lots of ways, and I always seem to associate it with the darkness and cold of winter.

At the end of December 1989, there was only enough to pay our wages for two weeks, and the council of the Royal Zoological Society of Ireland was talking about winding up the zoo. As the new year began, we were at the point of closure. I was only 23, but I thought my career at the zoo was over.

Judy and Kirsty – 2

In 2005, Judy and Kirsty were about to be moved to Germany's Neunkircher Zoo to join an elderly female elephant, Chiana, who had just lost her companion. We didn't have the facilities for training at the time, and their move was one of the things that made me realise how essential it was to start training elephants for travelling, rather than just sedating them.

In those days, you rang an elephant transporter who was skilled at moving elephants within the shortest time frame possible. Judy was a very headstrong elephant, and as she got older she became even more obstinate. The morning they were leaving I had to dart her. I went in, and we were looking at each other, and I'd swear she knew what was going to happen. When an elephant looks you in the eye, you feel like you're looking into their soul, and I felt so despondent, thinking, 'Why are we doing this? This isn't right.'

But I had to do it. It was a standing sedation, just enough sedative so that we could move Judy without her falling over, and it had to be done slowly. I needed to give her a couple of darts. She was letting roars out, and Kirsty, who was beside her, was getting upset. Kirsty had a sweet and gentle personality. She always wanted to be the keeper's friend and be close to me. And here I was, the guy who looked after them, upsetting her companion.

When Judy was in a standing sedation position, we had to get chains around her ankles. The truck had backed up to the house, and me and Ken Mackay, an excellent elephant keeper, got her under control. And that's exactly what it was: control. Judy had no choice. At every click of the chain, I kept thinking, 'This is so wrong.'

I was worried – not only about the physical consequences, because they were both getting on in age, but also about the psychological implications.

Judy was winched into the box slowly. You pull one leg forward, and then when the elephant realises it can't move anymore, it pushes the other leg forward. It is a horrible way to move such an intelligent animal. Judy went in unhappily, calling and screaming.

Kirsty was a clever elephant, and so calm that we didn't need sedation to put the bracelets on her legs. It was more like guiding a horse. I picked up the slack and was about to tie her onto the back of the winch machine to pull her in, but she felt the pressure of my hands on her leg and started moving forward herself. I just kept indicating left and right, and she followed my guidance. There was no force required, she was so compliant.

It was an emotional day, and a sad day. Judy and Kirsty were leaving the zoo after so many years with us, and it was a terrible way to send them off. It has been well documented that elephants handle stress not unlike humans. They can experience post-traumatic stress disorder. Judy and Kirsty must have felt a shocking sense of betrayal. I certainly felt that I'd let them down.

When they were gone, we demolished their crumbling bleak old house with its electric fencing and old moat, and started afresh. They also got a fresh start. Kirsty in particular was delighted with their new spacious outdoor habitat. Always curious, she was now keen to explore, and test, her new boundaries. By all accounts, with her skills at opening doors and fences, she kept the staff at her new home guessing! Joe Byrne was a regular visitor to see them.

The two elephants lived together until Judy died in 2011. The day before she passed away, reports described her as 'still alert and fit for her age'. In 2019, Kirsty made the news when she was reunited with the keeper who had looked after her at Calderpark Zoo in Glasgow in the 1980s. In 2021, surrounded by carers and companions, she was euthanised after a long battle with cancer.

SEVEN

ADVENTURES IN HAND-REARING

Snow leopards are fascinating cats. They live at very high altitudes, up to 18,000 feet, in the mountains of Central Asia, in countries such as China, Nepal, Pakistan, India, Russia and Mongolia.

They're very well adapted to their climate. They have a large nasal cavity that heats the cold air they breathe before it goes into their body. They have a lot of hair on their feet growing through the pads, which allows for traction on snow. Their tail, which they use for balance, is a wonder, and they have short front legs compared to their back ones, for running down rocky outcrops.

In the late 1970s and '80s, we had three snow leopards in the Lion House. Like other zoos around the world, we had been trying to breed them for years, but the cubs never survived. There were two problems: they were too closely bred genetically; and their house didn't support what they needed. Snow leopards need privacy and quiet, but their cage at Dublin Zoo was very open, and there were people around them all the time. So, while cubs were born nearly every year, they would either die or else the inexperienced females would eat them. They would start eating the umbilicus and continue eating.

In the 1980s, we managed to get a couple of cubs to survive, and we put an appeal out for a whelping dog to feed them. Sally, a beautiful golden Labrador, came into the zoo, and we put the two snow leopard cubs on her with a couple of her pups. One died soon after; the other kept suckling, but after about 20 days its eyes still hadn't opened. We were concerned. The zoo brought in an eye specialist, Dr Terry Grimes from University College Dublin, and he could see the cub had no eyes. They hadn't formed due to genetic weaknesses, and the animal had to be euthanised. However, Sally stayed on. She lived in Pets' Corner for years afterwards and was always a favourite with staff and visitors. She was a beautiful dog, with a gentle disposition.

In 1988, three cubs were born; two died, but one male cub, Patrice, battled on. When his mother abandoned him, Da, who was now head keeper, brought him home. We hand-reared him in Blanchardstown between us. He lived in an incubator at first and needed feeding every two to three hours. Hand-rearing an animal is incredibly tiring. It takes over your life. It was great to have the support of each other when we did hand-rear, and when my brother James came into the zoo, there were two sons available for the rotation.

Snow leopards are extremely difficult to hand-rear because of their immune system. In the wild, they don't really need one because bacteria and viruses don't tend to survive at the temperatures and altitudes at which they live. One of the big zoos in the US (San Diego, I think) spent a fortune on a large, state-of-the-art sterile area to try and hand-rear their snow leopards. Patrice was reared in a small family house in West Dublin, with the washing machine left on to get him to go to sleep (he loved the vibration). He would be driven to the zoo every day in a cat carrier or underneath my jacket. It was a huge amount of work, but in the end he survived.

Patrice lived at home with us for quite a while, and he loved to play with Dino. He was so agile that he would jump around the house and bounce off the ceiling and back walls. He grew into a magnificent, healthy adult cat. We introduced him to a female, Sasha, and we were hoping that they would have their own cubs. However, I returned from holiday in March 1997 to be told that Patrice had died of botulism from eating chicken while I was away. It was devastating to lose such a beautiful creature, the first hand-raised snow leopard ever in the history of the zoo — and probably one of the few hand-raised snow leopards anywhere in the world.

We kept learning, and over the years the snow leopards' conditions and habitats improved, along with the rest of the zoo. Finally, in 1993, Seena, who had come to us from the Bronx Zoo, had three cubs that she reared successfully herself – the first snow leopard cubs in the history of the zoo that were mother-reared. And it was just because the facilities were better: they had more space, a proper cubbing den and access in and out at night instead of being locked up. We were feeding them differently. I was the cat keeper at the time, and we made many small husbandry changes that I believe made the difference.

However, we still had heartbreak. Many years later, when my daughter was small, I took home another cub to hand-rear, from a first-time mother who wasn't producing milk. But despite nursing her day and night, she didn't make it.

Over the decades, we also hand-reared wolves, lions, chimps and lots of other primates. The apes were much more demanding, like having a child. For hygiene reasons, they had to wear nappies. They also had to be kept warm. I'd be going to the toilet with a chimp on my back because they didn't want to be put down.

Tyson was a lovely lion-tailed macaque, a beautiful ground-dwelling primate only found in a small area in southern India. Tyson's mother died two weeks after he was born in 1987, so I decided I'd take him home. He was tiny (less than six inches tall and only a couple of pounds) and needed a little bit of support. He came everywhere with me. In the wintertime, I'd have a big coat and he'd sit in the pocket for comfort. I'd be at the supermarket or bringing my mum to bingo in the car, and Tyson would be with me. I couldn't leave him down. He wanted to be with me everywhere. Eventually, however, I managed to reintegrate him back into the troop.

At the time, I was mad into boxing, and Mike Tyson was after coming on the scene. He was an amazing athlete, so I called Tyson after him without putting much thought into it. I got into awful trouble. Kader Asmal from the Irish Anti-Apartheid Movement said the zoo should have never allowed me to call the monkey Tyson, and said it was an example of 'appalling insensitivity'. There was a big storm in the papers, and I was called a racist. But it was the most innocent thing ever. I did it purely because Mike Tyson was my idol at the time.

In 1987, after a five-year pause, the zoo began breeding lions again. Unfortunately, lioness Marie rejected her two cubs, a male and a female. They came to my house in an incubator and soon graduated to a playpen in the kitchen, and then to kennels beside Dino. A public competition named them Socrates and Sheeba, and before the year ended they had both moved on to UK safari parks.

On New Year's Eve 1987 – the eve of Dublin's Millennium year – four lion cubs were born. One of them was very weak and small. She was being pushed out a little bit, so I asked Da if I could take her to hand-rear. She became Millie, the millennium lion cub. I'd often be out on the green in Blanchardstown with a lion cub on one lead and Dino on the other.

Millie became one of the zoo's symbols. She appeared on television, and I brought her all over Dublin, to functions and events. Looking back, I'm disappointed with myself for using her as a prop. That was the way the zoo got publicity and generated money in those days, and it was seen as a way of taking a bit of the zoo out into the community. But I didn't understand the damage I was doing to her: I was over-socialising her, and it was wrong. A few months later, Millie was moved to Longleat Safari Park.

Some hand-raised animals became very attached. While I don't want to be anthropomorphic, many animals – like humans – have the need to be social, the need for touch and recognition. Their responses are like ours. You can see the emotional upset, the crying, the need for comfort and all the emotions that we also have. We saw it with many animals that lived with us.

Apes, in particular, would really want to engage with you. Jamie was a character. He was a chimpanzee reared between a few keepers because it was such intense work – a 24/7 job. Ron Willis, who was head curator at the time, would take Jamie for a week, then my father and I would take him for a week, and then it was the turn of Helen Clarke. We'd be ringing around saying, 'I'm going out tonight, can you take the baby?' We'd be transferring Jamie from one house to another, with cots and nappies and Milton and all the bottles. It was quite a task.

I loved the opportunity to have young animals in the house, but it was without fully understanding the consequences of what we were doing. A lot of hand-reared animals become dysfunctional, live on their own or don't successfully reproduce. They don't have the body language or behavioural repertoire that would have been learned from others of their species.

Jamie didn't go back into his own troop at Dublin Zoo. Eventually, he was introduced to a nursery troop in Antwerp, Belgium. However, even though he was older, he was always very low ranking. I went to see him years later, and he definitely recognised me. He came over and sat down beside me; we looked at each other and had a moment. I felt like I had let him down. I could see he was living on the edges of the group. Chimp society is all about dominance. They get into these mad crescendos, with the dominant male showing his power over low-ranking males. It is not necessarily violent, but he will hit or push the subordinate members of the troop. Sitting with Jamie, I realised I was witnessing an example of how hand-rearing doesn't work unless there's a well-thought-out plan for reintroduction.

In 1992, Grace was the first gorilla born in Ireland. Her mother was Lena, who went on to have many other offspring in the zoo. Da confirmed that Lena was expecting using a human pregnancy testing kit. He used a syringe to collect some urine from a shallow recess, about the size of a small saucer, in her sleeping quarters and then squeezed that over the test. Polo was a massive gorilla that we had brought in from Germany as a mate for Lena. With his striking physique and presence, he was a magnificent specimen – the first silverback gorilla that we had seen in full size. It was like looking at a bodybuilder. Before Polo arrived, there had been two or three young males in with Lena who were being transferred to Singapore Zoo. Lena gave birth about seven months after they left. Since the gestation period for a gorilla is eight to eight and a half months, we figured that Polo couldn't be the dad. Arti, one of the younger males, had impregnated her before leaving. When the baby was born, Polo lashed out. We don't know whether it was because he knew the baby wasn't his (which is not beyond the bounds of possibility) or because the baby squealing upset him. Whatever the cause, he gave baby Grace an injury to the side of her head.

My dad brought her to Harcourt Street Children's Hospital, where she was diagnosed with a fractured skull. Grace came home with us to Blanchardstown. To have a western lowland gorilla living in a housing estate in Dublin 15 was not usual. Mam bottle-fed, bathed and oiled her (which she loved), and we'd all take turns minding her. However, she was so attached to Da that she would push the grandkids off the settee if they were near him.

Grace slept in a cot and wore babygrows and vests – not as a gimmick, to make her look like a real baby, but because she needed to be kept warm. If she had been with her mother, she would have been constantly held in her arms against her chest. She would also hold on to her clothes, or a teddy bear, for security, particularly when she was travelling, or Mam would swaddle her tight. She travelled everywhere with my parents, even when they were bringing my sister Mags, who was expecting a baby, into the Rotunda Hospital. Da stayed outside with Grace wrapped in a blanket. A woman in labour got an awful shock when she came out of an ambulance and saw him with a baby gorilla. Quick as a flash, Da reassured her: 'Don't worry, missus. I didn't get it in there!'

Some primates are so close to humans that, when something is wrong, a doctor can have better insight than a vet. Dr Ríona Mulcahy helped my dad with Grace and visited her regularly. She even wrote a paper comparing the development of an infant baby with a baby ape. It showed that the gorilla is way ahead, up to the age of five months, and then the human baby overtakes it when it starts to develop speech. A baby orangutan was named Ríona in her honour. She was the first orangutan born in Ireland, and she also spent about a year in Blanchardstown with us when she was rejected by her mother Leonie – another redhead in the house!

There's some great RTÉ archive footage of Grace on *The Late Late Show* in 1993. They showed some footage of my parents looking after her in our house in Blanchardstown, then Peter Wilson brought her into the studio for a chat with Gay Byrne. It was a great promotion for the zoo because people related so much to a baby primate. They were fascinated that she was being minded and fed just like a human baby. It evoked an emotional response in people, and they wanted to see her for themselves. The zoo capitalised on it because we needed to. Peter knew that attracting visitors in to see Grace was key to the zoo's survival, so he became a regular on *The Late Late Show*. Throughout my childhood and early career, much of the publicity for the zoo centred around its hand-reared animals. Animals that our family, and other keepers, reared at home regularly appeared in newspapers, on TV shows and at promotional events.

Grace was a gorgeous little character. She came into the zoo every day and became a favourite with visitors. However, zoos around the world were learning that hand-raised gorillas were being attacked when they were put back into troops. Stuttgart Zoo had set up a nursery, taking gorillas before they were a year old to teach them social interaction. Da took Grace there and stayed with her for a little while so that she could gradually become acclimatised (although I think he had to leave his jumper behind as a security blanket). She went on to be integrated with a troop in Germany's Wuppertal Zoo and did very well. Her daughter Vana returned to Dublin.

Saying goodbye to Grace was hard on our family because we had all become very close to her – we loved her. Animals are therapeutic. We're interlinked so closely with them that they can make us feel good. Hand-rearing animals made the bond even closer. It was wonderful to be able to share our world with these animals. However, while it may have been beneficial for the humans, we didn't realise the consequences for the animals.

Hand-rearing is something I now deeply regret. I really thought I was doing good and doing right, and so did other keepers. We just wanted to keep animals alive. It also had its benefits in terms of the zoo's profile, because we had an animal to pull out if somebody important visited. We would also often bring them to schools and old folks' homes to give people opportunities to get close to a wild animal. It's a world apart from what we do now, but at the time we did not understand the long-term ramifications for the animal.

Hand-rearing animals was the norm in zoos across the world at the time, not just Dublin. It was only later that people realised the complications. Because they had been reared with humans, these animals would lose their communication skills as a species. Far too many hand-reared animals ended up on their own or in situations where they could never be reintroduced. Or, if they were successfully introduced, they would tend to live on the periphery of the troop or pride, but never be fully accepted. It was a bit of a bankrupt life because of human interference. Even when they seemed to integrate, there were after-effects. Chimpanzee or gorilla behaviour had been imprinted by human behaviour. Or big cats would ignore the cats they were with and would want to come over to the fence to get a scratch from a keeper.

As a zookeeper, I don't want to see animals dying, but I have to think of the bigger implications of human intervention. As our understanding of animals has grown, we realise that inexperienced mothers losing litters of cubs, or first babies, is what happens in the wild, and they are learning all the time from the experience. In addition, when an animal becomes reproductively active, it can change its position within a herd or a troop or a group. For example, a dominant male gorilla can bully and push a young female gorilla out of the way. However, once she becomes reproductively active, she has something to offer the troop, and something to offer him, and immediately there is a change in his behaviour. This is just nature's way of securing the future of the troop. Her ranking and position change, even if she is too young or inexperienced to be able to rear a baby.

Even though hand-rearing is a lovely experience for humans, and even though we are doing it with the best intentions – to save a life – the consequences have to be fully understood. When organised structured breeding groups began in the European Association of Zoos and Aquaria (EAZA) and the British and Irish Association of Zoos and Aquariums (BIAZA), we started to understand the consequences, and there was a change in philosophy.

In the cat world, it is also not unusual for females to lose their first or second litter, because they can become fertile extremely young in zoos, probably because of high-nutrition diets. We were finding that, across Europe, hand-reared cats were socially non-reproductive or, if they were reproductive, they didn't know how to look after their babies. Hand-rearing was creating a vicious circle where animals didn't develop the correct behavioural, emotional and vocalisation skills that they needed.

By the early 2000s, the studbook keepers said that, unless it was a very rare case and there was a high probability of a re-introduction into their social group, we were not to hand-raise. The advice now is to let nature take its course. However, while it might be the right thing to do, it is one of the most emotionally harrowing parts of the job. I often went home in tears because we had to watch young animals die.

So we tried assisted feeding, where we provide supplementary feeds while keeping animals with their mothers, and that has been successful. In 2014, when Asiatic lion Sita gave birth, it was evident by the screaming of the cubs and their behaviour that they weren't getting enough milk, even though she was showing all the maternal instincts. So we trained Sita to leave her cubs and go outside for a few moments a couple of times a day. We would feed her, and the cubs also got supplementary feeds.

The key difference was that the cubs were being left with their mother. They weren't imprinting on us. Sita would come back in and clean off the milk that we had just fed out of the cubs' mouths, then she would be happy to nurse. The cubs would continue to try and suckle. Assisted feeding is now seen as the way forward, and it can even be done with apes. There have been cases where young females have good maternal instincts but low milk production, and they can be trained to come over and be fed themselves while their baby gets supplementary feeds.

In Dublin and all over Europe, there were animals that were the darlings of the zoo, the cute babies who were hand-raised by the keeper at home. I did it because, at the time, I felt it was the right thing to do. However, it resulted in animals that lived for a long time (apes can live up to 50 years) without ever properly integrating socially.

There was one final, but successful, hand-raising experience in my career – but that's another story!

EIGHT

TRANSFORMATION IN THE 1990S

Ashortage of money meant that director Peter Wilson could not make many of the changes he knew were needed at the zoo. As the 1990s began, we continued – rightly – to get a lot of criticism for keeping monkeys, apes, lions and other animals behind bars. There was a growing, necessary, public debate on animal welfare. Zoocheck and the ISPCA were among our critics, and the DSPCA asked the zoo to stop keeping elephants following the death of Judy, an African elephant in our care. There was also growing concern within the animal team about the well-being of our animals and their conditions, and constant calls for improvements in their facilities.

Mick Doyle, a former rugby coach and a vet, was appointed head of a working group to investigate the zoo, including how it was run, and the conditions and care of the animals. I think his brief from Taoiseach Charles Haughey at the time was to fix the place or close it. It was good that he was an animal man at heart. We all got a chance to speak to him, even a young keeper like me.

We all knew we couldn't continue as we were. Morale was very poor in the zoo, and it looked like we would be closing after the St Patrick's Day weekend in March. Peter launched the Save the Zoo Pound Appeal, and he also did a whip-round of the board members to raise £32,000 to keep the zoo going. It was a pretty depressing time. However, when the public was asked if they wanted a zoo in the city, the answer was a resounding yes. People were fond of their zoo, but they also knew it wasn't acceptable to keep animals in poor conditions.

The working committee found huge problems that needed to be addressed. In 1990, speaking to the Irish press after submitting the report, Mick Doyle insisted the zoo in its present form should not get a penny. He called the zoo structure 'cockeyed, with Victorian attitudes and management ideas'. New animal welfare standards and procedures needed to be introduced immediately, he said, and the zoo moved to a wildlife park structure and was taken over by a new organisation. The report was also critical of the lack of legal standards for keeping animals in captivity, calling for new animal welfare legislation that would also control circuses and pet shops. However, the report also described the zoo as 'a vital national asset' and recommended that it be given immediate funding and extra land to deal with structural problems. The government stepped in with a £5 million injection.

Everyone agreed we could not continue with the zoo in its old historical format, with animals pacing up and down, and apes and monkeys in barren cages. A £15 million development fund was approved in 1993, and the transformation began. Infrastructure was overhauled, and animal enclosures were improved. The first improvement was the long-overdue installation of a sewer ring. Until then, waste (and there was a lot of it) from enclosures including the llamas, blackbucks, orangutans and lions had drained directly into the lake!

We were on the verge of change, but most of our exhibits still had structural and husbandry problems – in other words, issues with how we were caring for, managing and breeding animals. However, there was an ambition to embrace a wellness-inspired philosophy, to provide the animals with better, more appropriate places to live.

There were also initiatives to improve keeper knowledge and modernise how we cared for animals. What took my career to a whole new level was being sent to the National Zoo in Washington, DC, on a four-month keeper exchange in 1992.

We were a 30-acre zoo at the time, and a large chunk of that was taken up by a lake, while the National Zoo covered 163 acres in the heart of Washington, DC. It's a free-entry zoo because it's part of the Smithsonian Institution, and it received massive resources – back then, it was over $100 million a year. Money was poured into creating habitats. Amazonia was just one multimillion-dollar exhibit. It showed all the different levels or tiers of a rainforest, and I had never seen anything like it. Coming in at the bottom level, I saw fish from South American rivers, including piranhas. As I moved to ground level, I could see agoutis and other rodents and mammals from the forest floor;

and then on to arboreal primates like golden lion tamarins and white-faced saki monkeys within the canopies; and right up at the top, all the various bird species. It was humid. I was immersed in trees and roots and vines. There were even rain showers. I thought it was just fantastic.

I got to work with some incredible species in the Small Mammal House, like tiger quolls, bizarre marsupials and little carnivorous animals I hadn't even heard of. I also worked with the elephants. In those days, their house was old, and there was a bigger focus on the inside areas because of the harsh winters they would get in Washington. We were still going in with the elephants then, in free contact, and the keepers were remarking on how good I was with the animals, how natural I was with them. That was because, with the environment and facilities in Dublin, we had to build up and maintain a relationship of stability and understanding with the animals.

The National Zoo had another location in Front Royal, Virginia, where they had an endangered breeding centre. That was the first time I had seen conservation in action, and what it meant. I got to work at the Cheetah Conservation Station, where endangered cheetahs were working on runs, chasing a motorised line that went around their habitat to mimic hunting. Their food was hung on it, and the cheetahs had to capture it and 'make a kill' to eat, giving them the opportunity to run at their natural speed of up to 70mph.

But while it was very progressive and modern, the National Zoo wasn't as good as Dublin in other ways. The people in one department didn't know people from another, so I became a link between them all. On St Patrick's Day, I brought about 50 keepers to an Irish bar called Nanny O'Brien's for a drink, and I remember people talking to each other for the first time: 'Oh, you work at the Elephant House' or 'You work at the Australia section …' So we set up a little weekly get-together. I was responsible for all the departments starting to connect with one another.

It was an incredible learning experience – not only as a zookeeper but also because it was my first time away from home for a long period. It really broadened my outlook and horizons. I was learning every day, taking notes, writing down ideas and thinking about what I wanted to take back to Dublin Zoo. 'We have to think more about the animals' became a mantra for me. My experience at Washington Zoo left a huge impression on my thought process.

I came back bursting with ideas. Dublin Zoo had some golden lion tamarins, which are a very rare endangered species – their population in the Brazilian rainforest was down to less than 1,000 animals because these small primates were losing out to massive deforestation. I asked if we could move two from the South American House to a larger outdoor area. I had brought back some pictures from Washington, so we started a three-dimensional approach to designing their quarters. I got hemp rope and wrapped moss around it, tying it with string to make it look like hanging vines. We wet it every day and put in bark and different formations, which were changed regularly. The behavioural changes we began to see in the tamarins were significant. It was with these small mammals that I first realised that a keeper had to become the architect of the animals' experience.

Some of the other keepers would be saying, 'Who does that little f***** think he is?' if I tried to do something different, but I was lucky enough to work in the section where Da was the team leader. He always encouraged and supported me. However, he was also great at making me recognise potential hazards to an animal. If I wanted to put ropes in, he got me to evaluate first if an animal might get entangled. That's what we do now when we're building habitats: we decide what natural behaviours we are trying to encourage and design around that, but we always have to assess risk first.

I got support for other ideas, too. I changed some of the primates' habitats, introducing substrates and mosses for the little monkeys to forage through for insects. We covered the back of the chimp area in deep bark chippings and sprinkled it with nuts and other foods. They spent hours foraging. We introduced bark for the lions to lie on. We were starting to get some ideas of what animals needed.

What I had learned in Washington also helped with our breeding programmes. I had wanted to move the tamarins because I believed we could increase their potential to breed in bigger open-air conditions. We had been giving them fruits and some monkey chow or tamarin cake (which contained their nutritional needs and supplements) every morning, but they would eat all the fruit and leave the chow or cake. We started giving them the most nutritious food in the morning when they would eat anything because they were hungry. Feeds of fruit were then given during the day. It is always great when lions, tigers and other big animals are born, but to see the first golden lion tamarins born in Dublin Zoo was very special, and that was a result of the skills and ideas I had taken back from Washington.

The golden lion tamarins are among the great successes of the zoo. Dublin has now sent many of these beautiful monkeys back to their native habitat. Those chosen for a reintroduction programme were sent first to a free-range area at the National Zoo in Washington, and then to Brazil for full release. At one point, we held and managed the international studbook for these endangered species, helping to decide which tamarins from different zoos would go back to the rainforest. Dublin Zoo has played a very significant role in managing the population that now exists both in zoos and in the wild.

If you walk through the South American House now, you will see how square cages with straight walls have been broken up with strategic design and planting. The tamarins, like many monkeys, are incredibly private animals that live in dense forests, so they have been given the psychological comfort of something that resembles their natural habitat, where they feel safe and secure because they think they cannot be seen. They can hide behind branches, or escape outdoors or into nest boxes. Their habitat triggers natural behaviours like scent marking, socialising, grooming, foraging and hunting for insects. Hot spots mimic the sun coming through the canopy, and irrigation systems help to replicate the humidity and the density of the rainforest from where these animals come.

An incredible range of habitats was completed between 1996 and 1998. One of the first to be created was the primate islands, to replace the old Monkey House and give primates increased opportunities outdoors. The chimps Wendy, Betty and Judy had been with us since the 1960s. As tea party chimps, they had been humanised – dressed in children's clothes and encouraged to mimic human behaviour. I had looked after them in the old chimps' pit, where they were locked into cell-like structures at night, with no stimulation and huddled up together to keep warm, because the only heat came from a little blow heater. I cried when I saw them going out on their island for the first time. They now had heated houses and comfortable beds, but they also finally had a place where they could feel the rain and the wind and decide whether they wanted to stay in or go out.

For the first time, the chimps had freedom of expression. They were able to climb trees, pick leaves and buds, break grasses, find insects... all the things they would be doing in the wild. They would lie back on the grass, or sit with their feet in the water on a warm day, picking and eating succulent plants. We had turned a corner, and there was no going back.

In 1997 I became a leader for the team that cared for apes, carnivores and, of course, elephants. I was directly responsible for a group of about 12 people, and I reported directly to my father, who was now the senior curator for animals.

That was also the year that the polar bears were moved from their pit to a purpose-built habitat with grass, trees and a filtered pool. Da even implemented an environmental enrichment programme, which included laying down scent trails with spices, fish and meat across their habitat, and the reward of a chicken at the end. However, the damage had been done, and we were all relieved in 2003 when they were sent to a far more appropriate zoo in Hungary, where they had separate male and female quarters and could come together when they wanted. I drove with them down to Rosslare. We stopped a few times on the way to check on them, and looking in the back of their crates I was happy for them and happy to see them go.

Aside from state money, Dublin Zoo also received land to expand. In addition to animals, joggers, polo grounds and cricket games, the Phoenix Park is where you'll find the US Ambassador's Residence and Áras an Uachtaráin, the home of Ireland's president. The government agreed that 18 hectares of the land around the residence should come to the zoo, including woodland and a lake, and incoming President Mary McAleese approved the grant in 1997.

The extra land was the making of the zoo. It was a mature, beautiful landscape with a stunning backdrop, and it gave us an opportunity to create new habitats and move all our large African animals. This would become the African Plains, which opened in 2000. There were consultations with the public and experts from around the world, but the inspiration for the design of these new habitats came from nature and the wild, based on what these animals needed to thrive. We created a mixed-species habitat with white rhino, giraffe, zebra, ostrich and oryx. The latter had been extinct in the wild at one point, and they are now being reintroduced back from zoo-bred populations.

These majestic animals were now sharing a habitat and showing their full range of behaviour. Visitors could marvel at large herbivores roaming plains and open spaces. We also gave the animals areas to retreat. I remember going up to the African Plains when it opened and people saying, 'They're too far away now, you can't see them' when for years they were giving out because the animals didn't have enough space! So we had to encourage the visitors to be patient.

My education continued on the ground. I was always eager to learn – I still learn something new every day. From 2001, I was sent to Sparsholt College in Hampshire a few times a year to study for the Advanced National Certificate in Management of Zoo Animals. This was a course run by Andy Beer, an amazing man who is now on the board of Dublin Zoo. The idea was to combine science with practical keeping, to give keepers additional status and skills so they weren't just seen as labourers. Sparsholt still runs what is seen as the gold-standard zookeeper course, teaching nutrition, habitat maintenance, design and record keeping.

Peter Wilson did a remarkable job keeping the zoo afloat through some of the most turbulent times in its history. He reversed its fortunes and began its transformation. It was also under his directorship that our uniform changed from the prison-officer look to a more casual safari-ranger style. This change was driven by staff, who wanted functional work clothes: having a chimp try to grab your tie when in full uniform was just one example of how impractical it was.

In 2001, when Peter stepped down, he handed over a zoo that was financially viable and was creeping its way back into the hearts of Irish society. Leo Oosterweghel, a Dutchman, took over as director. He had come from Melbourne Zoo with extensive global and visitor experience. He had worked his way up from keeper to director, so he understood what keepers wanted and needed. Along with new assistant director Paul O'Donohoe, who had worked with him in Australia, Leo had a great vision for Dublin. This included continuing to develop habitats that supported and promoted the animals' natural instincts and social groupings.

They started putting money into creating restaurants and food outlets around the zoo and installing proper toilet facilities. At the time I was disgruntled, thinking we should use this money for the animals, not realising there was a strategy. Leo understood that a visitor coming to the zoo with their child needs to be able to feed that child and change its nappy in comfort.

The long-term plan was to get the infrastructure right, get the appearance right, get the animal welfare right and get the keepers out to other zoos to bring back new ideas. Some of the most significant changes in the zoo came under Leo's leadership.

And gradually the culture changed to one where animals come first and visitors second.

Leo was opposed to giving animals anything artificial to interact with, like car tyres. He used to say, 'You may as well give them the whole car.' He was right, and this is now something I preach everywhere I go, particularly in America, where they use a huge amount of synthetic stuff. I'll be walking around a zoo with a director and I'll say, 'Look into this habitat: you have this beautiful landscape with beautiful animals, and you've got a tyre or a plastic drum hanging down, so the first thing a visitor notices is fakeness.' Zoos are under siege in a lot of places (and quite rightly so), and one of the reasons for keeping animals is education. However, if you want to educate, and inspire love and care for animals, you can't have a child walk away thinking that a gorilla plays with a tyre. Show them picking plants, eating flowers or wading through water, and you show the animals the way they should be seen.

Even how we named our animals changed. The job was given to the public, but the names had to have their origins in the animal's native countries. Leo also explained the importance of changing the language around animals in zoos as their environments changed: 'habitat' instead of 'enclosure'; 'in human care' rather than 'captivity'. And he was a stickler for insisting on how animals were depicted in photographs. He wanted quality pictures that showed respect for the animal. He would never allow a picture to go out to the press where an animal looked confined or behind bars unless it was a training zone, and absolutely never with a prop. He insisted we had a duty to represent animals in a respectful way, and the zoo regularly commissioned photographer Patrick Bolger, an empathetic portrait maker.

Leo brought in a Seattle-based company called Jones and Jones, which had built some amazing zoo habitats in America. All the team leaders were given a seat at the table to discuss what we wanted for the animals and to contribute to what was happening. We knew we had a moral responsibility to continue improving their lives. A master plan was drawn up for new dynamic habitats based on the animals' biology and the wild, where visitors would never again look down on the animals with disrespect.

People started coming back in droves, and it was a great era. Profits were being pumped back into the zoo, and everything started to change. The design and management teams would select a big project, and work would start in September when the season wound down a bit. It created a great philosophy where people would say that the zoo was constantly improving. It was fantastic to witness the habitats changing, but it would be 2007 before the opening of our most ambitious new habitat: the Kaziranga Forest Trail.

Bernhardine

The calf of Irma and Ramon, Bernhardine was born in 1984 in Rotterdam. She was named after Prince Bernhard, the husband of Queen Juliana.

As a young calf, she was sent to Germany for free-contact training, which would have been quite forceful. She quickly learned that if she did not submit to humans, there was a painful consequence. She then returned to the Netherlands.

As she got older, Bernhardine became aggressive. She made attempts to attack keepers, and it nearly had tragic consequences. To my mind, it was her way of saying she had had enough. She wasn't prepared to accept human dominance or the training methods that were used at that time anymore. She didn't like the consequence of the hook, and she rebelled against it.

Bernhardine earned the reputation of being one of the most

dangerous elephants in Europe. They stopped going in with her in Rotterdam for safety reasons, and she was kept on her own a lot of the time. But she was a product of her situation: there is no such thing as a bad elephant, it is all down to how they are managed.

Despite Bernhardine's reputation, we were keen to bring her to start our new herd in Dublin with her sister Yasmin and niece Anak. We wanted to take her because we knew we were going into protected contact, so we weren't going to be sharing the same space with her. Also, she was a sister and was pregnant, so we knew there was potential there. We predicted that, if we used the right methods, Bernhardine would probably flourish – and she did. She is now a truly remarkable elephant.

I spent a few weeks in the Netherlands working with our new herd before bringing them home. It broke my heart to watch her submit to being chained up, just so she could join the other elephants for a while. I understood that she had to be managed for safety reasons, but it convinced me that this type of management belonged in the past.

I cannot describe the joy I felt the day Bernhardine (Dina) went out into her new habitat in Dublin. Her confinement was over. She was unchained – physically and mentally. And from that day, the day people stopped going into her world and dominating her, Dina excelled.

We began a programme of positive reinforcement training with her. She was given the choice to walk away whenever she wanted, but she never did. When she came to the protected-contact wall for the first time, she had a choice: 'Do I participate or not?' It took a while to get her confidence, but she very soon understood that every time she followed us and touched a target stick, she got a reward. She quickly realised, since she was no longer in free contact, there were no negative consequences from the humans around her.

Her behaviour was a revelation to us. I remember talking with Donal Lynch (a first-class elephant keeper who was always thinking about ways to improve the herd's life) about our awe at the transformation we were seeing. Dina evolved into a wonderful matriarch for Dublin's herd: relaxed, calm and kind. She was the first up every morning, looking for interaction. Smart and quick to learn, she loved the reward-based training system. She bought into working with us because there were only positive consequences from being in the presence of myself and the other elephant keepers, who all worked so passionately to help her. Anytime she engaged with us, she got rewarded for it. She got jackpots. She got pleasure. And it revolutionised her life. She would come every day looking to be trained, looking for her hot washes, her pedicures and to have her tail creamed. She just loved it.

When we walked into their house in the morning, Dina would come over to the wall and drop down to her knees – a legacy from her history of submission. Then she would put her face up to the protected-contact wall and make these beautiful, glorious, harmonious rumbles that seemed to be a welcome, an acknowledgement of our arrival.

It was such a privilege to watch Dina transform into an elephant with autonomy and a full repertoire of natural behaviours, a wonderful mother and a matriarch. She had had calves in Rotterdam, but they didn't survive. When she came to Dublin, we gave her the conditions to be a mother and stepped away. We gave her a sand floor to give birth on, comfort, food, a dark, warm house and her family with her. Everything was ready.

She tapped right into her innate nature, and her calf was born without human interference. Dina gently pushed it into standing, as she, Yasmin and Anak trumpeted, rumbled and squeaked a welcome.

NINE

LIVING WITH LUCY

I loved being team leader for the primates, carnivores and elephants. I'd always seen it as the most exciting section of the zoo.

Apes are problem solvers and need constant stimulation. Orangutans, chimpanzees and gorillas all have spindle neurons in their brains, like humans, dolphins and whales (and, of course, elephants), and they have the ability to learn and develop throughout their life. It is thought that the chimpanzee, apart from man, is the only animal that understands the concept of murder. They form militias of five or six males, and they will target and kill a male from a neighbouring troop, mutilating him, pulling off his genitals and almost celebrating the death.

It's vicious behaviour, but it also shows how intricate their brain processes are.

Betty is a 61-year-old chimpanzee in Dublin. She and her companion Wendy arrived here in early 1965 from a sanctuary for orphaned chimps in Sierra Leone. Along with Judy, another West African chimp, they were the chimpanzees that used to do the tea parties on the lawn in the 1960s. Wendy passed away a couple of years back, and I'm dreading the day that Betty dies. When I would go up to her habitat, even if I hadn't seen her in a couple of weeks, she would give me a really warm greeting, stretching her belly for me to rub it. It always made me think about how lucky I was to spend so much time with chimpanzees. However, it also made me realise that even 60 years after being hand-reared, Betty still looked for human interaction, ignoring other chimpanzees in the troop.

I met my wife Leona when she was working in the Cabra House pub, my local. There's always been a bit of confusion about where I'm from. I was born beside Manor Street and moved to Blanchardstown, but many people think I'm a Cabra man. I've always spent a lot of time around Cabra. I had friends from school who lived there, and it was near the Phoenix Park and Manor Street. So when Leona and I started going out together, her home, with her late mum Josie and sister Erica, became my home. And I'm still there!

Leona has a great sense of humour and is great fun to be around. Luckily, she has always loved animals, too, because when we started our family in 2003, it wasn't with a baby – it was with a chimp.

Lucy's mum Mandy was an inexperienced first-time mother. In the animal kingdom, a lot of animals can lose their first litter or baby. One reason is that, when females start to cycle, they get mated by the male, but they're not necessarily mature. Mandy was quite young when she conceived, and she didn't have the skills to look after Lucy. She wasn't feeding, and the baby was getting weaker. She was putting her upside down, lifting her, dropping her, putting her down. The baby was screaming, and it was causing distress within the troop. We became worried that the dominant male Kongola might take it out on the baby because he didn't like the troop being upset.

Lucy was getting weaker and weaker, and I knew she would have to be taken out if she was to survive. By now a lot of zoos were against hand-rearing, but I said to Leo, 'Please, can we hand-rear Lucy? I've thought about it, and I think I can get her back into the troop. I live close by, and I'll bring her back into the zoo every day.'

A human family is not dissimilar to a chimp troop: there's a social structure, people coming and going, and lots of interaction. However, I had learned a lot from the time we had hand-reared Jamie, and I wanted to make sure that Lucy would be reintegrated successfully as an active member of the chimp troop.

We had to sedate poor Mandy, who wasn't doing well, and take Lucy away. She was a tiny little thing, only about a kilo and a half. It's hard not to be anthropomorphic when you see this vulnerable, defenceless little thing that just wants to be cuddled.

Lucy triggered Leona's maternal instincts, and when Lucy wasn't in the zoo, the two of us reared her at home. To be honest, Leona, her mother and her sister did more of the rearing than me. They spent hours with her. They always opened up their home to the animals I was taking care of.

It was just like having a hairy baby in the house. Lucy needed to be fed every couple of hours. She needed nappies. She needed the love and affection that human babies need. She slept in a cot at the end of the bed, and she'd be in the bed with the two of us when we woke up. She gravitated towards affection and warmth, exactly like a human baby.

She had the same discomforts as a human baby, too. On one occasion, I had to go to the 24-hour chemist on O'Connell Street at 3 a.m. looking for teething gel. The woman behind the counter asked me how old the baby was. You can imagine the look of surprise on her face when I explained that it was actually for a chimp. We had paediatricians looking after her because chimps are closer to a human baby than what a vet might be familiar with, and, when Lucy was older, she even attended a human dentist for some repair work.

She was just adorable, and such a little fighter. She travelled in a baby seat in the back of the car. When I'd pull up at traffic lights, people would be taken aback when they looked in the window.

Lucy displayed all the emotional ranges that you would see in a human baby. She would be happy. She would be sad. She would scream or cry if she didn't get what she wanted. She would latch on to us when she wanted a hug. We couldn't just put her down and walk away. She had to be with us. If I walked away suddenly, she'd start screaming, and raise her hands in the air as if to say, 'Please, don't go'. She would laugh hysterically when I tickled her around the neck and played with her.

Her behaviour emphasised even more to me the importance of ensuring the fundamental wellness of animals, creating an environment where they have autonomy, opportunity, choice and control over their own lives, and where they can express themselves. We have got to fully understand the consequences of not giving them a meaningful life.

It was hard not to over-humanise Lucy when she was living at home, but the most important thing was to ensure she understood chimp language. Chimpanzee politics are very complex and dramatic, so she needed to learn about them. I'd get her in to witness the sights and sounds and vocalisations of the chimps every day. Within weeks of being born, she was clinging on to the inside of my jacket, her little head sticking out, as I was feeding the chimps breakfast. They'd holler and howl when they saw her, and she started to hoot back at them, to give out to them. As she got older, she would be on my back as I fed them, and the other chimps would be bouncing around, making noise. Lucy would be holding on to me, but she was learning from them all the time. If they had a tantrum, or if they were playing, she'd hear it.

She thrived and became a real character. She was so funny. One day I could see all these workmen on the street looking at our upstairs windows and pointing – Lucy was banging at the window and making faces at them! She was just loved by the community. I'd pull up outside the house with her, and there would be kids in the garden waiting for her, and she'd get all excited upon seeing them. Or a couple of the kids would knock to ask if they could play with her. I'd let them go out the back or come in the house, and they'd be wrestling with her or running up and down the stairs, and we could hear her laughing. She was one of the gang. These interactions with the local kids were so natural. In a way, it replicated the play and interaction of chimpanzee society.

Lucy became a real favourite at the zoo, and people followed her life story. I had her on *The Late Late Show* and *The Den* with

Zig and Zag. She had a huge fan club. People would come up and leave gifts for her. I'd take her to schools and educate the kids about conservation and chimpanzees, so she was a great little ambassador for her species, too. She wasn't just a prop: while she was with us, she was bringing an awareness of chimpanzee society, and how they care for and look after one another.

While Lucy was living with us, our daughter Mia was born. It was like watching two children together, playing and pushing and shoving. One time Mia had a lollipop, and Lucy came over, took it out of her mouth and put it into her mouth. Mia grabbed it back and put it into her mouth, and they went back and forth like this for ages. It gave me further appreciation of how close we are to these animals.

Lucy's emotional attachment and understanding were phenomenal. I'd be lying on the couch and I'd joke with her, letting on to cry, and she'd run up, pull my hands off my face and hug me. It's horrifying to think that people kill chimps, dissect them and use their hands as ashtrays, or eat them, as they do in the bushmeat trade. Years ago, they would go out, shoot the mother and take the baby away to sell as a pet. This still goes on in some parts of the world. Chimpanzees have been found as pets in Europe and America. But chimps are not pets. We treated Lucy as a chimpanzee and tried to provide the social group she needed as a baby.

Lucy taught us so much. It was amazing to be in her world and to have her in our world. However, no matter how much we all loved her, I was very conscious from day one that the object was to get her back into her own family. We built a little nursery, a play area, for when she was in the zoo, by partitioning off the chimp house so there was a dividing frame between her and the others. She was in their area but not in the same space. I started to notice little positive interactions. I'd be going by, and she'd be up at the fence line and they'd be looking at her. As she got older, about four or five months, she'd be hitting off the mesh calling the other chimps to engage them and, over time, I could see that they were starting to try to groom her. She would turn her back and push it up against the barrier to be groomed by her mother.

I knew we were on the way to a successful reintroduction. Lucy started alternating spending her nights in the zoo and with us. Finally, just before she was three, we put her back in permanently with her mother, her grandmother and another chimpanzee. It was like she had lived with them all her life. Lucy gravitated straight back to her mother, who was the first to grab her. Mandy cradled and carried her, even though she wasn't feeding her. She knew Lucy was her daughter.

Of course, it was sad to say goodbye, but we had to put our feelings aside. Animals have great magnetism, and we want to be near them, but they are not meant to be in our world. The problem is there are chimpanzees all over the world that have been hand-reared by humans, and when they become adults, they lead sad and solitary lives stuck in cages on their own. They get too big to be handled safely, and they are left with a human void in their lives, without the repertoire of behaviour to succeed within their species. This is why appropriate habitats, diets and social structures are all paramount to optimise animal wellness.

Often when hand-reared animals are reintroduced, they are killed by their peers because they don't have the knowledge to be around their own species. They don't have the language they need, and they won't be accepted into the troop. But with Lucy, I had invested the time to make sure she understood chimpanzee behaviour. Seeing Lucy's successful reintroduction was one of my best moments as a zookeeper. She now lives a meaningful, fruitful life in Warsaw Zoo in Poland. She's there with her mother, her grandmother and another chimp to set up a new breeding programme. I'm hoping to go and visit her soon.

Aside from Lucy, my kids have always had an affinity with animals and have shown great skills with them. We've always had pets. Mia always wanted three of everything: three snakes, three turtles, three canaries, three geckos... I asked her once why she wanted three of everything, and she said, 'Because if one dies, the other two will always have a friend.' Sound logic from a child.

The smells and the sounds of the zoo felt like security to me – they still do. And they have the same effect on my own children. When I'm doing pedicures on the elephants' feet there's a very strong musty smell. My jumper and clothes might still smell of it when I get home, and Mia still says it reminds her of being a kid and how I smelled coming home from work. She might even ask me to take off my hoodie so she can wear it, almost like a security blanket.

I want to make sure they have a solid respect and under-standing of nature, and what it means to be kind to animals. It's very important for young children to be around animals to help their emotions to develop and teach them the responsibility of caring for another being. For a time, Lucy was a key member of our family, and I like to think that she taught us a lot about how to live together as a troop.

TEN

DUBLIN'S ELEPHANTS

My education in modern elephant care began in 2005 with Alan Roocroft, and it continues to this day. Alan, whose passion for elephants is extraordinary, has been my guide and teacher, and an incredible mentor.

We came into elephant care through similar pathways, beginning as trainee keepers in our teens. Alan, who is from Manchester originally, started at Chester Zoo and went from Hamburg to San Diego. He has seen, and worked, the brutality of the old-style management, training with a hook. He now introduces and advises on protected-contact management around the world. He has even been called in to look after Pablo Escobar's animals and Michael Jackson's elephants.

In 2005, Dublin Zoo entered a new era of elephant care. The old Elephant House was demolished, and we agreed that we should take nothing from the past. For over 100 years elephants had just survived in the zoo – they had never thrived. It was time to revolutionise elephant care, to come up with a new philosophy with elephant well-being at its core. We wanted to give any animals in our care the opportunity to live a genuine authentic life, on their terms, with choices and opportunities.

As part of the EAZA's European Endangered Species Programme, it was recommended that three elephants – Bernhardine, Yasmin and Anak – would come to Dublin from Rotterdam Zoo in the Netherlands to form the nucleus of a new breeding group. Rotterdam had reached capacity and needed to split their herd. It was decided to design an entirely new elephant habitat that would be suitable to keep a small herd of cows and a bull.

With government support, over €4.5 million was invested in a new 8,000 square metre elephant habitat. The anatomy of the elephant had all the answers for the design of the facility that we needed, and we used Kaziranga National Park in India as our inspiration.

We came up with a list of what we would want for these elephants. Alan, Leo, Paul, Da and I put a plan together alongside Jones & Jones to create a modern protected-contact programme that would centre on an elephant's biological, physiological and psychological needs. This meant that humans would not share the same space with the elephants. We would create opportunities for them to have choice, control and decision-making over their day, but we would train them, using positive reinforcement, to accept veterinary interventions and essential foot care.

The project took over five years to complete. The first phase involved the construction of the house and a small outdoor 'kraal' area, followed by an outside habitat of nearly two acres, with pools, feed poles, scratching rocks and sand topography that could be constantly changed. Elephant biology and physiology dictated that we needed long, undulating landscapes to walk around, natural light in the house, natural pools for swimming and strategic rocks, so no one animal could dominate the landscape, and there was always a safe way out of the water.

The three elephants arrived in Dublin in October 2006, and the Kaziranga Forest Trail opened to the public in 2007. Bernhardine and Yasmin were sisters, and Anak was Yasmin's first calf. The dynamics were perfect for establishing a herd. Social grouping and relationships are the key components to elephants thriving in human care. They have to have that investment in one another to create a meaningful life. They love being together and caring for one another. Even their feeding and sleeping repertoires are community-based.

Bernhardine (Dina), the matriarch, had been considered a difficult and dangerous elephant in free contact, but she quickly thrived in our protected-contact environment. Both she and Yasmin arrived pregnant, which was a great way to start our breeding programme. Elephants have the longest pregnancy or gestation period of all mammals, 22 months, so we knew we had some time to prepare. She wasn't due for over six months.

One substance was key to the success of our elephants' new home: sand. Historically, elephants have always been kept on hard surfaces – concrete, rubber and other substances that could be easily cleaned. Dublin became one of the first zoos in the world to use sand, both indoors and outdoors. The elephants were no longer living and sleeping on a hard, unyielding substance. It was a total game changer and a world apart from the cold, wet floors that Dublin's elephants had endured.

Their inside space had a seven-foot-deep sand floor. The type of sand is critically important, and we used sand from a river estuary. The particles are round, so it's like the elephants are sleeping on a bed of marbles. The sand fits with the contour of their bodies, sort of like a memory foam mattress. It is fluffed up every day, washed and irrigated. It epitomises comfort. We create these massive sand pillows, and most of the elephants sleep in a big pile on top of one another. It is a beautiful sight.

Elephants sleep only a few hours a night (older elephants sleep less than the younger ones), and they have to keep moving their weight around because they're so heavy. We know that our elephants' sleep is comparable to that of wild elephants because my colleague Brendan Walsh spent nearly two years doing a sleep study. This also showed the importance of choice: no elephant ever chose to sleep on concrete; they all went for sand. Some preferred sleeping near doorways, others under a heater; some liked the mounds, while others opted for a flat surface. We were able to change the topography of the indoor habitat to prepare it for when they decided to come in at night.

Outside, deep sand spreads as the elephant's foot sinks in when walking, making their muscles work harder. We use a digger to stop it from becoming compacted. Keeping the sand soft makes it even more comfortable underfoot. The benefits of sand for the elephants' feet cannot be overemphasised. Deep sand provides a natural wearing down of nails and footpads. The foot health of the herd is a stark contrast to the horrific foot problems experienced by Dublin's previous elephant inhabitants, and so many elephants around the world.

Asian elephants also love the interaction of sand and other substrates like bark and pine shavings. It allows them to fulfil their natural tendencies to dig and dust themselves and others after swimming or washing. However, one of the greatest benefits of the sand floor has been the help and support it gives during and after births. In the past, elephants giving birth on concrete would be extremely stressed because their calf would try to get to its feet, slip and fall. On concrete, it sometimes took 30 minutes or longer for a newborn to get to its feet. Sand, however, quickly absorbs the fluids produced during the birth process and allows calves to gain a much quicker grip when trying to stand. Thick sand also cushions those first falls. Watching our first calf, Asha, being born in 2007, we were amazed to see how quickly the amniotic fluid was absorbed into the sand and how quickly she was up and walking.

The next seven of Dublin's calves were fathered by Upali, a proven breeder with wonderful social skills, who joined the herd in 2012. The complexity of elephants is fascinating. They communicate with each other through their faeces, using chemicals to transmit a lot of information, a bit like a dung message: 'Hi, I'm a bull with high testosterone,' or 'Hello from an older female in oestrus and ready to mate. I'm nearby.' We arranged an olfactory introduction with Chester Zoo: we sent them some of the cows' faeces, and they sent us some of Upali's. Our females now knew there was a new bull in town, and they would probably meet him soon.

It made for a very successful introduction. When he finally arrived, Upali was greeted with delight. I don't think we had ever heard such a range and volume of vocalisations from the herd. They closed ranks, ears flapping. They were all touching each other for support, trunks intertwining and passing on messages of comfort. Bernhardine, the matriarch, broke away from the herd to welcome him, backing into him with her tail out. It was her way of telling the others that everything was okay. Yasmin, her sister, was a bit more excitable and ran around quite a bit. Upali just followed her until she stopped, put his trunk on her and calmed her down completely.

Twenty-three months later, Yasmin gave birth to her second male calf. It was a long labour for her, but Kavi was up and walking when he was seven minutes old. A month later, this was followed by the birth of male calf Ashoka to Anak and then female calf Samiya in September 2014 to Dina. These were the first elephants to be bred and born in Dublin. It was a huge moment in elephant global conservation. Dina oversaw everything, and they all offered support and help to one another.

Upali's habitat had its own pool and hay hoist. He could drift in and out of herd life – and he usually chose to head back to his own space every afternoon. It was usually a bachelor pad, but the younger females and males loved having a sleepover there when one of the females was in oestrus (faecal hormone tracking is used to determine the oestrus cycle and to establish fertility). Young bulls, from the time they are born, are destined to leave the herd, and being around the bull, they are witnesses. They are learning the behaviours required to be successful bulls and breeders from the sights, sounds, smells and hormonal responses.

When the Kaziranga Forest Trail was first envisaged, our plan was to create a self-sustaining, multigenerational zoo herd in protected contact that would contribute to the international breeding programme for endangered Asian elephants. Any elephant that leaves Dublin Zoo now has its own repertoire of elephant behaviour that they can tap into, and that will make them successful breeding males or females. Budi, one of Yasmin's calves, has already moved to Denver; Kavi and Ashoka to Sydney; and Yasmin and Anak will be taking their sons to Cincinnati to establish a multigenerational herd and bring new elephant bloodlines to North America.

Another aspect at the core of Dublin's success has been giving the herd the choice of being inside or outside whenever they want and providing constant access to food. In the past, Dublin's elephants were fed once or twice a day, but when we looked at how elephants feed in the wild, it became apparent that they spend up to 18 hours a day searching for food and eating. A 24-hour feeding strategy allows the herd to feed like wild elephants, with different foods distributed around the indoor and outdoor habitats. If you watch a herd feeding, you'll see that nobody is rushing. Nobody is worried about controlling the food or getting there first. They cooperate with each other.

The elephant brain also needs challenges. They don't want easy feeding. They don't want to feed off the ground. Those thousands of trunk muscles need to pull and tear and stretch, so we created high feeding opportunities using hanging feed nets filled with hay, maize, browse or feed balls. These are on hoists, which are operated on rotations, and raised and lowered by remote control. A hay net might drop at one end of the habitat, and the elephants walk and feed, then another is remotely activated somewhere else. The elephants could pull that net down if they wanted, but they never do. They prefer to weave the hay out. They love those natural prolonged feeding patterns. It's far more

enjoyable for them. We pack the net with nuts, fruits and vege-
tables and the older elephant hits it and knocks down food for
the calves to eat while she feeds in peace – she's creating a jackpot.
Everybody's getting fed, and everybody's enjoying the experience.
This elevated feeding also triggered a remarkable increase in the
neck and shoulder strength of the herd.

Browse – high branches – is of great importance to elephants.
They eat huge quantities of it and need access to it on a continuous
basis. It is very high in fibre but very low in nutritional value.
We'd usually plant and wet at least 20 pieces before the animals
came out in the morning – mini forests they can forage and feed
in. Their houses are also redecorated daily with planted browse, so
they have access through the night. They pick the sweet pieces off,
break off the top buds, and then come back later and knock the
bark off the stick, which is exactly what they're programmed to do.

Elephants are happy to go foraging through bark chippings,
breaking branches or digging. We have pipes filled with food that
they have to bend down on their knees to get into, improving the
natural flexibility in their joints. We hide fruit under piles of pine
shavings or bark. Root balls from a tree are packed with fruit and
nuts. These stimulate pushing, shaking and all kinds of physicality.
Food is buried up to a metre deep in the sand floor of their house.
Elephants use their trunks like mine detectors, and when they
realise there may be some fruit or a root vegetable underground,
they dig and have to displace maybe a couple of tonnes of sand
to find the turnip. They've also learned that it tastes better when
they rub the sand off on their leg, or when they go over to their
water trough and wash it!

Feeding options are rejuvenated throughout the day because
this is how elephants live. They go in search of opportunities.

The arrival of automated feeders brought a whole new dimension to animal care. There is now a remote-controlled system of feed pods. A text can be sent to a pod, and it throws out a scatter feed or releases some browse, before sending back a text saying the elephants have been fed. It can be done day or night. I've done this as I was flying across the Atlantic. At 39,000 feet I checked in on Upali and was able to feed him by triggering a feed pod to open. Feed walls are also timed to open throughout the night, with holes for different trunk sizes, so everybody has an opportunity to get in. The walls contain hay or feed balls. These balls are packed with hay, but they also contain nuts, and the elephants have to shake the nuts out. It's a great form of stimulation.

Technology has also allowed us to monitor them without interference. When I was operations manager, I'd give the keepers time to look through daily camera footage, to look back and see what had happened the night before, making sure that everyone was fed, everybody was happy and everybody slept. It's such a privilege to observe some of the hidden moments and to watch how they nurture and protect their young. I recall a particular moment captured on CCTV: the animals were inside in different parts of the house, but something (maybe a fox or a cat) came in or near the house, triggering an alarm call. The communication was remarkable: it looked like they were all being 'sucked' to the centre into a group, in a protective ring around the young. This powerful moment was an example of the cohesion of a multi-generational herd working together to protect the calves.

Observing the animals over a 24-hour period has informed important research, prompting adjustments in feeding routines and opportunities or changing habitat topographies. It also gives a greater understanding of the relationships, coalitions and bonds within the herd. This was key when making decisions about choosing elephants that will be moved elsewhere as part of the international breeding programme. The wrong decision could have a major impact on the elephants that get left behind, and those that leave.

The integrity of the herd is the core of everything. Every year we asked ourselves if there was something we could do differently. What did we need to re-evaluate? How could we improve life for the herd? There were limitations on space, but we realised that it's both the quality and the quantity of space that are important. A locomotion study mapped out every single square metre of their habitat, and we were able to establish areas that the elephants weren't using – because there was nothing to bring them there. Often there was a simple solution: putting a log pile in that corner or creating a feeding opportunity. We got to a point where our elephants were travelling similar distances to those in the wild: Upali was moving 15km per day, and the cows and calves more than 9km, and it was meaningful, purposeful movement.

Modern elephant keepers spend very little time with the elephants in terms of direct contact. They are architects of the environment. The team's job is to provide what the herd needs and to create opportunities for the elephants to show species-appropriate behaviour. When I looked after Kirsty and Judy, I would always think about what happened when I wasn't there. After going home, I'd worry about how I had locked them in, that their food would probably be finished in two hours, and that there was nothing else for them to do. That's something I emphasise to all young keepers: 'Remember when you go home at night, it doesn't get any better for the elephant, the lion or the tiger. The only choices and opportunities they'll have are what you've left them. That's how important your job is. There's no magic fairy popping in at five in the morning and doing something if a keeper hasn't done it.'

In Dublin, we created a scenario where it didn't matter when the zoo closed for the night. Now, when the keeper goes home, the house becomes alive – and not just because of the timed feeding opportunities. I could look in at night-time on the cameras from my phone and see the elephants foraging or swimming in the pool. I remember one time looking in during a really heavy thunderstorm. I was worried about them, thinking I should have locked them in. Instead, I could see six or seven elephants out in the middle of the habitat on a hill, looking up at the skies as the thunder and lightning beat all around them, and warm rain lashed down on their bodies. They could have stayed in their house, but they chose to experience the storm together. And then the thunder and lightning subsided and off they went together, as a herd, to swim in the pool. This episode epitomised to me what our elephant programme was about: elephants being elephants.

Wherever the herd goes out in their habitat, they know they're going to be hugely stimulated. The keepers make constant landscape changes. Inside, the sand floor is turned and manicured daily to keep it fresh and create a stimulating surface; outside, the landscape is also changed daily. Mini forests spring up; new root balls are placed for the elephants to rip and tear for the sap inside; the sand is moved into hills and rises for them to climb and negotiate. Elephants' needs vary at different life stages. What the calves want to interact with is very different from a larger elephant. Creating a very large sand mount (up to three metres high) allows an older elephant to lay down much more easily because she doesn't have to drop so far into the ground. We're always trying to surprise them. A new day should bring a new experience. They love olfactory, sensory stimulation, and sometimes we'd cover areas with different scents like honey or lavender.

We also create mud wallows by digging holes with a machine and filling them with clay and water. A wallow uses about 10 tonnes of organic mineral clay, which comes from potter Nicholas Mosse in Kilkenny. It is rich in nutrients and really good for the skin, and the elephants love it because it's like a mud bath. The wallows are especially good during a heatwave when they allow the elephants to act as they would in the wild: wallow in oily mud to create a sunblock (with an inbuilt insect repellent) on their skin. The calves in particular love wallows, where they roll, wrestle and play. Elephants start to mature about 10–15 years, but they are still calves up to age seven or eight, and they want to play. All the calves born together play together like kids in a playground. They just want to be together, to eat together, to sleep together.

The success of what has been done can be seen in simple moments like watching a calf wander into a mud wallow, realise it's getting deep and let out a roar to show she's worried. Within seconds, her older sister comes over, touches her with her trunk and guides her back out. Interactions like this show love, care and guidance.

Water plays a huge role in the elephants' day – and not just the 40 gallons they might drink to stay hydrated. Just like in the wild, water is never too far away. They all get a daily hot wash and a cold shower, which they love. Young elephants love having the water directed into their trunk for spraying and drinking; it can be very calming. Strategically placed water cannons create immediate interaction and stimulate them to swim in the pool, which holds 300,000 litres. Though Asian elephants love to swim, elephants in a zoo environment will rarely take themselves into pools, unless they are playful young calves or under extreme weather conditions. A sprinkler system or water cannon is a great way to stimulate them to enter deep water. The pool has a very important social function, too: it's wonderful to watch the bonding that happens in and around the pools.

A small herd of Indian blackbucks share the elephants' outdoor habitat, and peacocks, pigeons and ducks frequently land in it or swim in the pool. Interacting with other species, maybe backing up when a blackbuck gets too close, is a natural behaviour.

The Kaziranga Forest Trail was designed to immerse visitors in the landscape and lifestyle of the Asian elephant. As you meander along the path, you hear birds singing, bamboo rattling (we grow 25 species) and water flowing, and you might see a peacock, or catch a glimpse of an elephant through the vegetation. If somebody just dropped you in there and asked you where it was, you would probably say somewhere in Asia. The immersion can feel so powerful. Everything planted on the trail by the horticulture team comes from Asia, including a Himalayan lily that only blooms every 10 years and that people travel to see.

There are strategic viewing areas where visitors can watch these majestic animals feeding, playing, digging, bathing, sand-dusting each other or just relaxing together. When the trail first opened to the public, it was wonderful to hear people telling their children to be quiet. We have moved from a culture where people were bringing bags of nuts or apples to feed or throw at the animals, to one of respect, where they watch with appreciation and deference as these beautiful creatures feed, play or swim. They're seeing these elephants do what they might be doing in the wild.

We showed the world that elephants can live and behave

like elephants while in human care. They have all these opportunities and only just drift in during the morning for us to give them their daily care. They no longer stand around waiting for something to happen. We shouldn't get a pat on the back for giving elephants what they need – it's the very least we can do if they are in our care. However, I don't underestimate the importance Dublin's elephants have had globally. Our elephants live an authentic life. It's not the wild, but we have created opportunities for them to express their natural behaviours, and we allow them to be elephants. We no longer dominate them, and we are now insignificant in their world.

The day we opened the habitat, I gathered all the keepers around and said, 'The journey only starts now.' Some of the faces in that team have changed over the years, but it has always been a wonderful group of elephant care professionals and some of the finest people I have ever worked with. They have included my brother James, Ciarán McMahon, Donal Lynch, Alice Cooper, Ray Menzel, John O'Connor, Ann Penny, Mel Sheridan, Rachel O'Sullivan, Hannah Wilson, Christina Murphy and Brendan Walsh.

And I'm proud to say that a lot of what we all did in Dublin Zoo is now outdated. We know we need to continue to push the boundaries, to move the remaining herd into the next era of elephant care.

Asha

For all its history with elephants, the Republic of Ireland's first elephant birth didn't happen until May 2007.

When Bernhardine (Dina) and Yasmin arrived from Rotterdam, they were both pregnant. We had a mating date for Dina, so we knew she was due first. The gestation period for an elephant is 22 months, though they have been known to deliver at 21 or 23. The reason for this long gestation is to ensure that the calf is 'well cooked' when it arrives, so it is ready to move with the herd within minutes. A newborn elephant can travel 15–20km on the first day of its life. The herd has to feed, so it has to move.

During pregnancy, elephants might put on over half a tonne in weight, and their stomachs will look huge, like a hay belly, when they eat a lot. During the latter stages, you might see the calf move or the mother-to-be might kick out a back leg, which shows she is in a little bit of discomfort.

In early May 2007, we were beginning to see behavioural changes in Dina, and our elephant consultant Alan Roocroft confirmed that labour was imminent. All the team watched on CCTV in my office: Donal Lynch, Alice Cooper (one of our first female elephant keepers and an excellent trainer) and my right-hand man, Ciaran McMahon, who would later share a cigar with Paul O'Donoghue to mark the occasion!

Alan, who was watching from San Diego, talked us through it on the phone, calming us down and reassuring us that it was all good, all natural and to just let them at it. We were on standby, but we wanted this most intimate of moments to happen without human interference.

It all went as nature intended, and I remember feeling ecstatic as we saw the new calf on the sand floor. If winning the Lotto brings people joy, this brought us ten times more. We watched in wonder as the three other elephants welcomed the calf, and we looked at each other and said, 'We've done it. We have helped create something special.' It was so exciting.

Anak was pushed in because she was the youngest member of the herd. She got to learn from the sights, smells and vocalisations that happen at that birthing moment, so she could tap into this information later on in life. Dina, Yasmin and Anak circled around the calf, inch-perfect with their feet, welcoming her.

Then, if we hadn't been warned, we might have started to panic. Dina started to kick her calf. It looked like it was violent, like she was kicking hard, but Alan told us this is what they do to stimulate the calf. They kick it on the back and put their legs over to push and shove it because they want to get the calf standing and moving. Their genetic blueprint tells them that there could be a predator nearby, so it is vital that the calf gets up quickly and moves.

The calf was standing within a few minutes. Dina snuggled and minded her newborn with a delicate intimacy. It was just beautiful to witness. Then she bent down on a knee to reach eye level with her calf, and the two gazed at each other. We were all in tears, recognising that the new start Dina had been given, and Dublin's new philosophy for elephants, had all culminated in this 110kg bundle: Asha.

ELEVEN

HOW TO TRAIN
AN ELEPHANT

How do you train an elephant? I work on the basis that every elephant is an individual, but I always use three key elements: target training, with positive reinforcement, in protected contact. Other important words are patience, calm and consistency.

Protected-contact management is a system where animals and human carers do not share the same unrestricted space. All the procedures that require being close to the animals – pedicures, health care, washing – are conducted through a protective barrier.

Protected contact was first developed in Asia to manage dangerous elephants. Wild-caught elephants were kept in a large enclosure – a corral – after capture. In the first stages of training, the trainers would approach the elephants from the outside of the wooden structure or bars, and present food or treats in order to try and gain the elephant's confidence.

In the 1990s, Alan Roocroft and marine animal trainer Gary Priest introduced and developed a modern protected-contact programme at San Diego Zoo, initially as a way of approaching the management of bull elephants. Alan would later teach me and the Dublin team how to target-train elephants, just like he has done in other zoos across the world.

We now design and use protected-contact walls with a series of ports, or openings, so that an elephant can put its foot or ear or tail out to us, into the human world, but we don't share the same space.

The EAZA has best practice guidelines for designing protected-contact walls, but the requirements will differ from herd to herd. The species of elephant will dictate how big access spaces should be for safety reasons. African elephants are far more flexible with their trunks, which can pass through extremely small areas. However, the elephants need to be able to see their trainers and recognise body cues.

It's all about building a positive association. The concept of positive reinforcement training for animals is linked to the work of behaviourist and psychologist B.F. Skinner. He considered many ways in which behaviour could be changed by treating someone or something differently based on what they did. His theory (that following a behaviour with pleasant consequences means that behaviour is likely to be repeated) is known as operant conditioning.

In positive reinforcement training, food is used to reward behaviours that we want to be repeated. Elephants do the behaviour – such as lifting their foot or presenting their ear – and they get the reward. It's a process that's also used with children, and it can be used to teach different behaviours or strengthen existing ones. The main principle is to reward positive behaviours. However, we also ignore the undesirable ones, standing back and denying treats. Every elephant has its favourite food, but mixing up options such as chopped apple, melon, sweet potato and carrots usually keeps the elephants interested in training.

Target training is the training of the animal to present a body part to touch a target, and it can be used for moving, stationing or gaining access to the animal. Elephants are on one side of a protected-contact wall, and the keepers on the other side get them to recognise the feel of a target stick (usually a stick with a small ball on top) when placed on their body. Initially, those sticks should be quite long, so trainers aren't standing too close to the training wall.

Training begins by presenting a target stick within the visual range of the animal so that they can actually see it in your hand, and then we offer food. Next, we present the target stick within a touchable range of the animal and feed them. We reinforce any movement, any curiosity towards the target, by feeding. Eventually, the elephant is in front of the keeper who touches its head with a stick and immediately throws a piece of apple on the ground. We would keep repeating that: a touch to the head, a feed to the floor. After eight or ten repetitions, the elephant starts to associate the stick touching its head with the reward. Then we move the target stick away slightly and reinforce any movement towards the target with a reward. The elephant will come with us – they are so intelligent.

Animal training is a very skilled practice, and there is no shortcut to experience. Trainers are teaching the elephant that touching the target brings a food reward. There is a real skill to capturing the exact desired behaviour and immediately reinforcing it with a food reward. Even the briefest of movements towards the target should be rewarded. However, that is not easy when you are concentrating on target sticks and feed bags, and elephants that are in constant motion. It is easy to mistime reinforcements in the early stages of the learning process.

We keep the training sessions short and finish before they get bored, to keep them keen. It's about the quality and not the quantity of time. If the elephants are unsettled and excitable, or if there is lots of play behaviour, it's better to end a session rather than persevere with unfocused animals. Giving elephants a little 'time out' allows them to settle down before starting the session again. When reward resources are removed from elephants, they tend to be much more focused when they reappear. Consistency and reinforcement in training lead to a longer concentration span over time.

Positive reinforcement training relies totally on the animal's natural problem-solving instincts, so it's also a great form of mental stimulation. They quickly learn that every time the stick touches a different body part and they present it, they get a piece of food.

Eventually, we introduce different body parts. We ask them if we can touch their foot, and we go through the same procedure. In the old days (and still in many places across the world), a hook would have been put at the back of the leg to get them to lift it for foot examination. Now, with the touch of the stick, they will lift and hold their feet out through an opening in the protected-contact wall for a lengthy pedicure. We take blood every week from their ears for testing. And it's all because they are being rewarded. You cannot spring any surprises on an elephant – we invest in time and training.

We train to do tusk trims. A broken tusk is like a broken tooth, and it can cause severe pain and infection. Filing and sawing tusks is a tough but necessary job. If our cows were coming into oestrus, we didn't want our bull to injure them, or damage his tusks, when he rested his head on them during mating. Upali's father in Zurich had massive tusks, but Upali had small, thin tusks. When Upali was mating a female, or jumping on their backs, he was in danger of fracturing them, so we trimmed them as a preventative measure.

We also cut down the tusks of our younger elephants to prevent damage when they are interacting or playing. In 2014, we had three elephants born within a couple of months of each other: Kavi, Ashoka and Samiya. To have three born so close together was very rare, and we were the only zoo in the world where it had happened at that time. They grew up together, played together and learned together. When Kavi and Ashoka would play with Samiya, they would spend a lot of time with their heads on her back pushing around her rear, but when they played together they would bang heads, preparing themselves for later in life when they might be fighting for breeding or territorial rights. We would trim off about half an inch of tusk, so they didn't break them. It also helped the tusks grow bigger and stronger. They have beautiful tusks now. Females have tushes, very small tusks, and Samiya did break one, but because of the training she allowed me to flush it out and clean it every day, so it didn't get infected.

We have also trained the elephants to allow us to flush their trunks if we need to take a sample for health reasons. Tuberculosis (TB) can decimate entire herds in the wild, but in zoos there are strict protocols, when an animal is being moved, to ensure that it is TB-free. To get a sample for TB testing, 100ml of saline is placed inside the trunk, and the elephant is taught to blow into a bag to get the sample. It needs to be trained to hold steady, lift the trunk in the air at the request of the trainer, bring the trunk down on request and allow the trunk to be shaken into the bag. The introduction of water into the trunk has to be accurate; we have to wait for the elephant to exhale before asking it to lift the trunk. Elephants will often begin playing games by holding back some breath and then blowing the water out, in order to get through the process more quickly, so it can take time to get it right.

Whenever we carry out even a small medical procedure, we try to make it a positive experience for an elephant. If we had to do a blood draw from a leg, we'd heat up the leg first with warm water to get the blood flowing, and then ask the elephant to present the leg into our world. One keeper will be at its leg, and another at its head feeding it and talking to it. A supportive strap would then be placed along the back of the knee to get the vein up. Before even considering approaching the elephant with a needle to draw blood, we would train it for pressure. We would count to three, and every time it heard 'Three', it would receive some melon as pressure is placed on its leg with a small screwdriver or similar implement. The animal realises that every time it feels that bit of pressure on the back of its leg, it gets rewarded with food up the other end. It is now associating the pressure with pleasure. Finally, the vet comes in and on the count of three, they insert the needle and draw the blood.

It's up to the trainers to create an atmosphere of positivity and to respond to the elephants' actions with immediate rewards, but everything happens on the elephants' terms. They always have the option to walk away. However, we've found they are happy to interact and engage with training or whatever medical intervention is required. They know good things happen. We have six and seven-tonne elephants allowing access for veterinary care because they're being rewarded for it.

Like humans, some animals will learn quicker than others, and I never expect immediate results. A lot will depend on how confident the elephants are, and how motivated. However, every elephant is different, behaviourally and physically. Some engage in the training process very quickly, while others take longer – it's like a classroom full of children, and everyone has their own way of adapting and learning.

I have to get to know the personalities of the elephants, to figure out each individual profile, and their peculiarities and preferences. I also have to establish the relationships within the herd, in order for training to be a success. Some of them will want to train next to one another, while others will prefer not to train next to a particular herd member.

Training is the only time that an elephant might be separated from the herd. As part of morning training sessions, they get individual keeper attention, rewards of their favourite foods and a full body wash – a hot wash followed by a cold shower. And then they love celebrating as they rejoin the herd. One of the benefits of individual training is that we can tailor diets for individuals within the herd and ensure that each elephant gets access to the nutritional material it needs. If there is only group feeding, a dominant elephant might eat quickly, and then push another elephant away to grab their food or treats.

In Dublin, training for calves begins when they're around seven or eight months old. We begin this young as part of our EEHV (elephant endotheliotropic herpesvirus) detection and prevention programme. This is a strain of the herpesvirus that can be fatal to both Asian and African calves in the wild and in human care. However, the cohesion of the herd is of utmost importance. They need to be together. We never close a door between a mother and calf. They will check in regularly with mammy, and mammy will check in with them. We just borrow a few moments of their life to inspect and train them for veterinary access and foot care, but the herd can always come together in a couple of seconds if needed.

As calves develop, they have a huge interest in food; and when they accept food, we get access to their mouths. By placing their trunks back on their head, we can take their temperature or do mouth inspections. The colour of an elephant's mouth is a good indicator of its health. Nowadays, to take an elephant's temperature you can point a laser into its mouth. You get them to salute (which means they lift their trunk), point in the laser gun and get a reading. We have calves running in to be weighed on the scales, lifting their trunks or letting us check their eyes – and they are relaxed. Their psychological wellness is paramount, and they know that they can go back to their mothers to check in, so they feel secure.

Individual stalls are used every morning during training sessions to separate the elephants and allow essential husbandry access. Calves can be trained in stalls adjoining their mothers. When leaving the inside habitat, all elephants are trained to walk through a large emergency restraint chute. The floor of the chute is a weighing scale, so we can weigh each elephant every day. This is also where we carry out medical procedures, give antibiotic care or take ultrasounds of pregnant elephants, so it's vital that they become accustomed to walking through it every day, stopping and lifting up their feet.

We prepare and train the elephants for any possible situation where they may need additional care. We don't want to spring any surprises that might stress them. With reward training, we also get them accustomed to accepting restraining bracelets on their legs, in case any invasive process, rectal examination or other veterinary procedure requires moving their legs apart or making them secure. It's important to get them used to being checked from behind and for rectal examinations since this is how we might have to administer fluids or antibiotics. It could be life-saving for them, so we need them to be relaxed enough to allow us to rehydrate them or give them medicine this way. It never seems to bother them – they're used to passing huge poos, so they don't seem to notice.

The most common injuries to elephants are bites to the tail or ear, and trunk, teeth or foot issues. This is why it's medically important to have access to an elephant's feet. Elephants are moving and searching for food for 18 hours a day, so their feet get quite a bit of wear and tear. A lot is going on in an elephant's foot that we can't see. There's a whole complex range of bones that go into each individual toe, and they have to be kept so that they can move freely. It's particularly important for pregnant elephants that they have optimum mobility and comfort.

Foot care is one of the most important parts of the management of elephants in human care. In the wild, their feet and nails get worn as they walk over a range of surfaces in search of food. In the zoo, with good nutrition, their nails grow quickly. Their feet are checked daily to ensure they are balanced, clean and free from debris, but every two to three weeks they might need to be heavily pared to maintain vitality. We would have one person working the elephant, rewarding them and keeping them on target, while a keeper or vet works on the feet.

In Dublin, the elephants' habitat has been designed to meet their biological and physiological needs, so they walk around on a range of different substrates, including sand and bark, similar to the natural substrates they would encounter in the wild. There is natural erosion, but the skin still does grow hard, and it can be uncomfortable. In order to give the elephants maximum comfort underfoot, their nails are regularly pared down. If a nail overgrows, it can compromise the integrity of the foot bones. The object is to keep the nails above the pad because if the nails are too long, they will hit the ground first, putting pressure on the cuticle and potentially leading to water pockets and abscesses. An elephant can spend so much time in the water they get 'pool cracks', so they may need filing to take the pressure off the nail, so the pads are hitting the ground first. If the nail hits the ground first the crack will develop more.

In older elephants, the nails and pads are much harder, so they can get overgrown, or stones can get caught in pockets. Excess growth on the foot can create a lot of problems, so feet have to be kept smooth. Not only do foot pads pick up sounds and feel vibrations, but they also help to pump blood back up into the body.

On the day before foot care, the elephant will be given a hot wash and a cold shower, and a Botanica herbal cream is rubbed into its cuticles and feet to soften them. We often joke that Dublin Zoo is the Hilton Hotel spa for elephants!

Pedicure skills are learned slowly. When I teach foot care in Dublin, I always bring the keepers to the elephant skeleton in the old Haughton House to see its intricate foot bones. An elephant's foot looks so different on the outside; you can just see the nails, the thick skin and the under part of the foot, a cushioned pad of fatty tissue that acts like a shock absorber. However, behind this pad – which has hard skin that needs to be pared – is a very complex foot and bone structure. Extreme care is needed when trimming pads and filing nails, or you could make a catastrophic mistake. When I'm looking at the foot, I know what's underneath, and that guides me to be careful. If I get it wrong, I will hurt the animal. Foot care is very intricate and requires a good understanding of an elephant's physiology.

A very sharp hoof knife is used to remove all the hard skin from the under part of the foot. A sharp knife is essential. If a blade is blunt, you will compensate by applying more pressure, and that is when slips and gouges can happen. There is no room for error, because the pad is quite soft, and cutting too deeply can result in long-lasting problems: not only can it cause permanent damage to the foot tissue, but you will also lose the elephant's trust. Luckily, it's colour coded: the overgrowth is quite dark while living tissue is light-coloured.

There should never be any evidence of blood after the foot has been trimmed or treated. The elephant's foot is formed in such a way that it is essentially walking on tiptoe, with a tough, fatty part of connective tissue for the sole. The exterior sole is ridged and pitted, which contributes to the sure-footedness of the elephant on a wide variety of terrain. The vast amount of pressure being exerted on the lower layers of the foot means that any damage can take a long time to heal and can often develop into a range of other long-term problems.

Training elephants also means training keepers. Keepers need to learn to understand individual elephants, their movements and their reactions. Interpreting body language comes with experience. I can look at an elephant and know that it needs a time-out.

These days, one of my roles as a consultant is to teach how to train elephants who need to be transported. The option we used years ago (and that many zoos still use) was to sedate the animals, and the risks were huge. After being given a mild sedation, an elephant would wake up with chains or ropes around its legs, and it would be winched into a box screaming and dragging. It is terrible to imagine a sentient, smart, intelligent animal being treated like that. Alternatively, it might wake up in a darkened crate. You can only imagine the trauma and the stress that caused, and the nervous wreck that would come out the other side. Thankfully, Dublin is one zoo that has led the way in training elephants for travelling.

One of the main things in any elephant transport is restraining the animal because an elephant that's free in a box can do quite a lot of damage to him- or herself if they can turn around. In addition, because they are such large animals, it's very dangerous (for both the driver and the animal) if they can displace their weight while being transported. So we train them to present their feet and allow us to put on a foot bracelet. Then, when they walk into the box, they present their foot at an opening or port, and we secure them safely into the box with a chain.

Crate training elephants (so that they will happily walk into their crate and remain there with the doors closed) can take months. Good zoos crate train, and I believe it should be a prerequisite for moving elephants and other animals because their psychological and physical wellness is at stake. I've seen lions and tigers come out of crates, and their noses are worn down, or there are red-raw patches on their head from rubbing and trying to get out of the box. If you train them, they will go into the crate; you feed them, they go to sleep, and they're happy to come out the other side. Training is a key component for successful animal welfare.

We knew when we opened the new elephant habitat that our work was only beginning, and I said to Leo, 'If we are successful in our breeding programme, we have to agree that any animal that leaves here has to go somewhere better, or somewhere that offers the same standard of care, and that includes not pulling elephants into crates screaming and roaring.' He just nodded at me, but I knew that meant he was in total agreement. Thirteen years later, two of our young bull elephants actually lay down in their crates and snoozed during their 30-hour transfer to Sydney. It was a sign of how comfortable and relaxed they were, and their mental wellness. We had prepared them well.

Everything that happens in any zoo now has to be done with the animals' well-being and welfare at the core. The EAZA has finally decided that the protected-contact management of elephants should be the sole management system of elephants within Europe by 2030 – but I don't think that's fast enough. No animal should be in fear of pain when they see a keeper approaching.

Budi

Budi, son of Yasmin, was the first bull elephant born in Dublin. When he arrived, on 17 February 2008, we were amazed to see him up and walking within seven minutes of his birth with the encouragement of his mother and other female relatives. It was a great moment.

The behaviour of the herd after his arrival was the subject of an academic paper published in *Zoo Biology*: 'Effect of a birth on the behaviour of a family group of Asian elephants (Elephas maximus) at Dublin Zoo'.

Budi was a playful calf, always active and full of fun, popular in the herd and with the keepers. He was also very smart from a young age, and quick to pick up new skills. This included realising that his growing tusks were very effective for ushering playmates in whatever direction he wanted them to go. As he got older, we noticed that some of the young female elephants had been unintentionally hurt by him during his boisterous play, and we predicted his behaviour would soon not be tolerated by the herd's females. There was also a risk that he could cause injuries.

We knew it was time for him to move on and leave the group, just as he would do in the wild. Young bulls are not allowed to stay within the family group. They leave the herd and form temporary bachelor groups. Then, when they mature sexually, they tend to live solitary lives and roam on their own. They will interact with elephants of all ages and sexes but only join a herd when there are females receptive to mating.

It is interesting to watch the body language of maturing bulls when they meet. They will puff themselves up and push their shoulders out when they're confronting one another. They stand high up on their front nails and elevate their ears out to the side of their head, a bit like fellas in nightclubs doing the show-off dance to make themselves look more impressive.

With Budi's best interest at heart, he was sent to Antwerp in Belgium to join two young bulls he could wrestle and play with, and then on to a bachelor group in Denver Zoo, Colorado. We trained him for the journey, and he walked into the crate calmly. We could see him looking out the window as it was lifted in the air, and we knew he would be fine.

In June 2013, Budi became the first Asian elephant imported to the US in more than 30 years. As he has no relations in North America, he will become an important male in the US breeding programme.

In Denver, he was given a new name, Billy, in honour of a long-time supporter of Denver Zoo, and he continues to be looked after well by the exceptional keepers there. He's still happily hanging out – and show-off dancing – with his bachelor buddies, in a habitat that includes two miles of trails, mud wallows, shade trees, scratching trees, full-immersion pools and swim channels.

I visited him twice in Denver when I was attending elephant care workshops. It was great to see him grow up and develop. He still hasn't reached full maturity, but he has turned into a magnificent bull, an impressive tusker. He has a calm disposition and a great relationship with the bulls that he lives with.

Watching him, I thought back to the morning he was born. And I was so happy to think that this Dublin-born calf is now about to take his place, an important place, in the North American breeding programme to conserve Asian elephants.

TWELVE

BRINGING THE WALL
TO CHINA

In 2009, my dad retired as senior curator for animals, and I took over from him with a new title: operations manager for the animals and grounds. I was also still the manager of the elephant programme. My morning routine began to include a check on all the habitats and catching up on the keepers' report sheets from the previous day. Then I'd make a list of what I'd need to talk to keepers about – a gorilla who hadn't fed overnight, a chimp with a cold or a behaviour issue that had been flagged. My day might include liaising with veterinary staff, deciding on nutrition or habitat changes, or teaching volunteers.

I relished the role because it was bigger, and I was working side by side with all the other departments. Keepers get a lot of recognition, but a good zoo is a combination of all the departments working and planning together: HR, marketing, catering, education, horticulture and volunteers. We'd all sit around a table and have our input, and I'd talk on behalf of the animals about what we needed.

One of the things I asked for was resources to train other animals using positive reinforcement. It was working very well with the elephants, and I was keen to spread this approach throughout the zoo. Our elephants were happy to give a blood sample, knowing they were getting a melon at the end of it, but I might still be firing a dart to sedate a chimp, or the orangutans Sibu and Leonie I had carried around on my back, so that we could take a sample. It wasn't fair.

We started training primates to present their shoulders for injections or blood draws beginning with the pressure of a blunt pencil. We also trained them to get on to a weighing scale. Weighing animals regularly is really important because when you're knocking them out for a medical procedure, the drug dose has to relate to their body weight. In the old days, it was guesswork and keeper experience. A lot of keepers became very good at guessing, including Da and me. The vets would often call me for a guesstimate before sedating a tiger. I'd have a good look and say, '120kg'. They would have mobile scales and would weigh the tiger at the end of the process, and it would usually be within a kilo of 120kg. There are now scales in the Carnivore Houses, and the big cats have been trained to stand on them, so every couple of months a record is taken of their weight to monitor their health and in case they need to be sedated for any reason. I even trained a tiger with a cardiac condition to allow us to listen to her heart with a stethoscope!

Training to get on a weight scale is also important for small animals like marmosets and tamarins. When we look at a creature that size, it's impossible to know if they have dropped 20g or 40g, but that would be significant weight loss in a 400g animal, so this is a great way of doing regular checks without putting them through the stress of being caught in a net by their keeper.

Positive reinforcement training is always non-threatening, and it eliminates any coercion or dominance of animals, while staff are kept safe. The modern approach is to teach animals of every shape and size behaviours that allow the provision of health or husbandry care, and for keepers to understand the principles of training.

We brought in a lot of experts to help us learn, including world-renowned training consultant Barbara Heidenreich, an amazing woman with an amazing understanding of animals, who believes that anything with a central nervous system, even an earthworm, can be trained. Susan Friedman, a psychology professor who's also an animal behaviourist, came to advise and guide us. I've worked with her since, and there is a great fusion between my intuition and practical skills with animals and her intelligence and scientific understanding of behaviours and responses. Despite our very different approaches, we always end up in the same square box. And this is where the future of animal care lies: with the Gerry Creightons and the Susan Friedmans marching together.

We introduced a philosophy of training for the right reason. People will remember the old sea lion show, where fish would be thrown to get them to jump onto a platform. Visitors loved it, but at its core, it was disrespectful to the sea lions. They are now trained only for veterinary access, so a modern presentation in

Dublin will show them presenting a flipper, where a blood draw would happen, getting their eyes or heart checked, or rolling and getting their head into a mask, which they would need to do if anaesthesia was required for a surgical procedure.

Technology has also brought massive changes. Cameras were first utilised in the 1970s, so that lion and tiger births could be watched on a monitor, but the first state-of-the-art CCTV system was installed in the new Elephant House in 2005. Technology advanced rapidly until I had an app on my phone that allowed me to check in on the herd any time of the day or night from anywhere in the world. I could also take photos or record videos if I noticed any interesting behaviours, and save them on my phone. By the time I was ops manager, there were 25 cameras around the zoo that I could access on my phone. The first thing I'd do when I woke up was look at the cameras, and the last thing I'd do before I went to bed was log on to them. Sometimes I'd even get up and check the zoo cameras during the night, particularly if a birth was imminent – I'd get an awful slagging from the rest of the team about it.

Seeing the new elephant habitat coming to life at night was powerful, and it resonated with us all. Why were we locking other animals in for 18 hours a day? We stopped shutting animals in, except for emergencies and habitat enhancements, and gave them back choice and control. Watching the rhinos crash playing with each other outside, big cats sleeping under the stars or the chimps emerging to forage as the sun rose over the park was such a joy to behold.

Historically Dublin Zoo had a reputation for breeding lions, but it was more by fluke than science. Our success with elephants was different. We were getting international recognition for what we had done – nine successful births over 10 years with no humans involved, only elephants teaching elephants – and my new role allowed me to get out and beat the drum even more. The zoo always put me forward to speak because they knew I would never be stuck for words, and because of my passion when I talk about animals, the zoo and what I want for the future of animals in human care. Alan Roocroft was also pushing me on, mentoring me and encouraging any new ideas I had for improving elephants' lives.

I can't overemphasise the influence that Alan has had on me. He invested time in me, bringing me to workshops – and then, eventually, to speak at workshops – in England, Europe and the US. When he began referencing Dublin's elephants and my team as good examples, I realised that we were doing a good job, and it was something to be very proud of because he doesn't suffer fools when it comes to elephants' welfare.

It doesn't matter where we are together – a coffee shop in Dublin, a bar in Arizona or having a sandwich in Hamburg – the conversation is always about elephants. His knowledge is just phenomenal. I like to think I'm a reasonably good communicator, but I probably learned a lot of that from Alan. He's a very charismatic man, and he has such a great way with people. I've been all over the world with him, and he will remember the names of everybody in hotels and restaurants, the people at reception, the people serving him food, the porters … people that he might not have seen in a year or two. And he goes out of his way to say hello. Other people might gravitate to the top table; Alan treats everyone with the same respect.

I continue to take inspiration from Alan. Like him, I love educating people about elephants and changing people's views and perceptions. I also love the variation in the audiences I talk to. I recall one day being in my son's crèche, helping three-year-olds fertilise a garden with elephant dung, and the next day I was up in Áras an Uachtaráin speaking to President Michael D. Higgins. Everybody wants to learn about elephants.

Whether it's three visitors walking around the zoo or 500 people in a room, I speak with the same enthusiasm and interest. I'm passionate about communicating about what we can do to help this world because it needs all the help it can get. I began to travel more. I would speak at the EAZA annual conference where over 600 zoos would be represented, and everybody wanted to know about our elephant programme, how we were getting it so right, and how they could change. Other zoos, like Oregon Zoo in the US, started to follow our design and our philosophy.

I was the ambassador, and I oversaw the programme, but our success was due to a wonderful team: management, the keepers, the vets, the education people, the horticultural people, the people who raked the sand and the people who cut the browse. Together we created something beautiful that resonated around the world.

I was spreading the word that there was a kinder, more respectful way to treat and manage elephants in human care. They did not need to be dominated. They didn't need to be standing on a concrete floor rocking from side to side when their food runs out after an hour. They needed constant choices and opportunities for purposeful movement, 24-hour access to their habitats, and the ability to decide what to do, whether it was going out or having a swim.

It got to a stage when we were getting so many requests for talks and conferences that Leo had to start saying no. I was ticked off because I wanted to spread the message, but he reminded me I was working for Dublin Zoo and I needed to be there. 'Get them to come here and see you,' he told me, and hundreds of them did. Everybody left saying they were going to make a difference to their own elephants.

One director who decided to change everything was Shen Zhijun from Hongshan Forest Zoo, in Nanjing. In China, elephants were still being kept in very small concrete enclosures and dominated by the hook, and he wanted to do better for his elephants. He came to see what we were doing in Dublin, which led to me going out there with Dr Carin Harrington from the Golden Bird Foundation, who has done a lot of good work advising Chinese zoos on modern design and animal care. We installed what I think was the very first protected-contact wall in China, and keepers quickly saw the benefit of the elephants being trained with positive reinforcement. Dr Shen also built a pool for Nanjing's elephants after seeing how much Dublin's herd loved soaking in the water and swimming.

I taught the keepers how to utilise the training wall, and how to do foot care. When I first arrived, the elephants had awful abscesses on their feet and overgrown pads. I taught the keepers about appropriate feeding and giving the elephants more exercise. It was very rewarding to see how they wanted to change for their elephants. Keepers started coming into Nanjing from Beijing, Shanghai and various other zoos. At one point I had a classroom of 30–40 keepers, all wanting to learn how to improve their elephants' lives.

I gave a talk about elephant care and wellness, and they came from all over China to listen. Dr Shen held a dinner for me. After all the serious talk, all the men became like little boys. When a huge fish was brought out, Dr Shen picked out the eyeball and put it in front of me. The interpreter told me that it was a great honour: 'Dr Shen has given you the eye of the fish to eat, and you have to eat it.' I couldn't be rude, so I swallowed it. Then they took out the Chinese wine and waited for me to drink it to see my reaction. It was worse than poitín, incredibly strong. The interpreter asked, 'Dr Shen wants to know what you think of his really strong Chinese wine,' and I replied, 'Oh, I thought it was water.' Everyone laughed, and he poured me another. It left me in bits, but it was fun and all part of the game.

I visited many zoos when I was there and believe I had a positive influence, making them re-evaluate their methods and their training. China's zoos are changing, but there is still a long way to go. Animals are still seen as props, and people will throw rubbish or fire bottles at them to wake them up. In some zoos, I saw very disturbing conditions. A bull elephant in Shanghai had been chained up for months because they were afraid he would break the structure. He was in a chronic stereotypical behaviour, rocking back and forth. It deeply saddened me, and I tried to advise them on what they could do. That's why I keep returning to China: to continue my mission to enhance the elephants' lives. Dina, Dublin's matriarch elephant, and I even feature on some of the signs in Nanjing. I also became an attraction in some rural villages. People had never seen a white person in the flesh – and certainly not a redhead – and they were fascinated, coming over to touch my skin and hair.

One of my most memorable visits was when Dr Shen took me to Xishuangbanna, a protected forest area near the border of Myanmar, where the remaining wild elephants of China live. China has only about 200 or 300 elephants in the wild. We stayed in the rainforest, and at night a massive python went past my cabin as fireflies were exploding outside my door. It was mesmerising. Even walking through the forest and hearing the different animal sounds was fantastic – the cicadas were so loud, I thought it was a chainsaw at first. When we got a call that the elephants had come down to the river, we ran. I got to see my very first wild elephants, a small family group that had come into the river to bathe, drink and feed off succulent plants at the water's edge. I was only about 20 metres away, and to see them in their natural home – to see the family unit together, the mother guiding her calf into the shallow water – was a powerful moment that will stay with me forever.

However, the park also had a show where elephants were riding bikes, kicking footballs and lifting tourists in the air. This was such a contrast to the way the wild elephants were living only a few hundred yards away. These performing elephants were facing a lifetime of suppression and dominance, and that was very difficult to come to terms with. Elephants continue to be kept for the entertainment of humans, and we need to spread the message that this should not be done and that we should not support this.

I don't know if the wild elephants I saw were among the group of elephants who captivated China, and the world, when they left Xishuangbanna in March 2020. Millions watched the drone footage as the animals trekked for thousands of kilometres across the country for 17 months before returning home. They were guided through villages, and across railways and roadways. Food was left out for them, and roads were blocked to try and redirect them. They ate rice crops everywhere they went, and they were seen knocking down doors to investigate houses and shops; they even managed to turn on a water tap with their trunks, lining up to drink.

It's not known why they decided to march, but there is a theory that there may have been food and water shortages in the reserve. Human expansion and rubber and tea plantations have all encroached on their habitat, and human–animal conflict continues to be a problem. Between 2013 and 2020, over 80 people were killed or injured by wild elephants in the province. However, their epic journey did seem to mark a cultural change in people's attitude to these wild creatures.

Meanwhile, back at home, we continued improving the lives and habitats of our other animals. In 2013 Dublin became one of the first zoos to stop the practice of pinioning birds (removing the last joint of one wing when they are chicks so they can't fly), and we created a massive aviary where the flamingos could flap their wings and fly across the pool. The saltwater pool in the new Sea Lion Cove allowed the sea lions to swim through different depths.

Our intent was to provide animals – from elephants right down to the smallest primates, the pygmy marmosets – with habitats that would encourage natural behaviour and provide the motivation for feeding, territory mining and socialisation.

More mixed habitats were introduced. The sloths live alongside the Goeldi's monkeys; red-footed tortoises feed on the food dropped by the saki monkeys; and siamang gibbons have joined the orangutans. White-crowned mangabeys live with the western lowland gorillas, and we have seen some fabulous interactions. You may now see baby gorillas and monkeys hugging each other and playing. The silverback won't bother with them, but the younger gorillas have a lovely relationship with their companions.

Before the Gorilla Rainforest habitat was opened in 2011, Lena and Harry (the silverback), and their youngsters, were living in what can only be described as a fish bowl. Surrounded by glass, they were on view constantly, to visitors and each other. Their outdoor area consisted of grass, a fixed climbing platform and a few branches. While plans began for a brand-new habitat, we installed trunks for them to climb and had full trees delivered so they could eat the browse throughout the day and have places to hide.

Their new island home has 15,000 plants, 200 species that they can eat, so they can forage throughout the year. They enjoy seasonal variations and pick and choose what they like best. It's a beautiful landscape. They have several huge trees that they climb and play on, and they now use their binocular vision to see down the Phoenix Park for up to five kilometres. The moat around their island keeps them back from the public, but it's also part of their environment, and they can interact with it. They can wade in up to their chests to pick plants. There's a netting in the water, so when the gorilla is in the water, it is using its feet to grip, as if it were climbing, so it's safe. There are mounds and private areas where they can get away from one another.

When we began to put out scatter feeds, we also noticed that they started spreading out over bigger areas, searching for food particles, and their vocalisations immediately increased. They were communicating more, letting each other know where they were and what they were doing. It was wonderful to see their behavioural repertoire change. They now act like wild gorillas. They tapped right back into their DNA, into how gorillas are meant to live – and all because they moved from a sterile environment to one that provided mental and physical stimulation. It was inspiring to see how a habitat enhanced natural behaviours.

The mental and physical condition of the zoo's apes has dramatically improved, as they climb trees and investigate their ever-changing habitat. They're interacting with water, chasing ducks off the island, catching birds and feeding off insects. We know it's not the wild, but we have come a long, long way. Dublin Zoo now stands with some of the best in the world, and Ireland's seasonal variations are a real plus. It's neither too hot nor too cold. Our animals are often outside 365 days a year. There are zoos in Europe where elephants have to be indoors for months because of the cold.

On my morning rounds, I'd always stop at each new habitat and say, 'Wow, haven't we done well?' Of course, we have to have access. We need to train the animals, to borrow some moments of their life for their well-being and healthcare, but we've stood back. And what we've found is that, the more we stood back, and the less we interfered in their world, the more appropriate behaviour we witnessed, particularly with our elephants and apes.

I remember coming around at seven o'clock one morning, and one of the young gorillas was at the top of a tree performing pirouettes as if she was doing ballet. All the others were in the house, but she had the confidence to come and dance high above the park, immersed in her own world, oblivious to the humans nearby. It was a beautiful sight. To me, she epitomised what we wanted for our animals. She was well and happy, and she was behaving as she would in the wild.

Upali

The bull who kickstarted Dublin's elephant breeding programme, Upali was a magnificent, charismatic animal, picked specifically because of his qualities of being a good dad – we wanted a calm bull who would contribute to the herd, who would be gentle and caring with the cows, and who would be a role model for his sons.

We were very familiar with Upali because he came from Chester Zoo. We've always had an excellent working relationship with Chester and share similar philosophies in animal care. We'd had our eye on Upali for several years before he arrived in 2012, and we built the bull house specifically for him. He came to us just before turning 18, as he was starting to come into his prime: five tonnes of masculinity and calmness. Standing three metres at the shoulder, he was a gentle giant.

Boxing at the annual Dublin vs London tournament in the Barbican Centre in London. The fights were always tough and competitive, but representing your county or country was always a great honour.

Me and Tyson, a lion-tailed macaque, whom I had hand-reared at home and introduced back into the zoo.

Introducing the bull sea lion to some of the new volunteers/tour guides at the zoo in the 1980s. (© *Dublin Zoo*)

The 1980s – two Patagonian cavies that I had hand-reared.
Hand-rearing was a lot more common then. Thankfully it doesn't
happen as much now. (© *Douglas Duggan*)

Me with Vicky and Sonny, two hand-reared chimpanzees, in the old chimp area, having my neck groomed by Vicky and playing with Sonny.

Sheeba and Socrates – two African lion cubs we hand-reared at home – being playful and jumping on top of me. (© *Douglas Duggan*)

The first mother-reared snow leopards born at Dublin Zoo. We had been trying for many years to breed a snow leopard without success, so this was a historical moment for the zoo and a very significant contribution towards conservation. (© *Mac Innes Photography*)

Sheeba the lioness, another lion we reared at home in Blanchardstown. She was a beautiful animal and very playful. Eventually, she went off to a safari park in the UK.

Alma – the beautiful male Siberian tiger who used to come to greet me every time I came near his habitat. A magnificent specimen and always liked to interact with his keepers.

(© *Jonathan Pratschke*)

Lucy at Leona's house (now our home in Cabra). Leona and her family took great care of Lucy and she and Leona were inseparable. Just after this picture was taken, Lucy managed to pull over the Christmas tree.

Lucy used to have a cot at the end of the bed but, like any baby, she needed comfort and touch, and we would regularly wake up with Lucy in the bed.

The ring-tailed lemur – always a real favourite. Here I have my nephew Anthony feeding a lemur. Anthony is now a keeper at the Zoo.

James and I with the two old elephants, Judy and Kirsty. The hook in my right hand was used for controlling the elephant. Thankfully, this method of training has long been confined to the history of Dublin Zoo. We now have a much more caring and kinder way of managing the elephants in protective contact using positive reinforcement.
(© *Colin Keegan/Collins Photo Agency*)

Mia with a little Asiatic lion cub who needed some additional feeding and support because its mother wasn't producing milk. Mia came up to the zoo and we were bringing it from surgery back over to its mother after some supplementary feeding. Mia has a great attitude and innate tenderness towards animals.

Zac and Mia in the dinosaur egg. A nice family moment.

Me with Alan Roocroft, a friend and mentor for over 20 years. Alan's enthusiasm and knowledge of elephants is incredible and he has inspired many elephant programmes and many keepers all across the world in creating better conditions for elephants and keepers.

Me with Gerry Senior after I had done a charity boxing match a couple of years back. My father has always been a great inspiration and mentor – one of Dublin Zoo's finest keepers.

Katie Taylor and Zac at Dublin Zoo. Katie kindly allowed Zac to wear her Olympic gold medal around his neck and Zac kindly allowed Katie to wear his first boxing medal. Katie has been a great supporter of the zoo and is very conscious and aware of conservation issues globally.

Family day at the zoo. This picture was taken for the Dublin Zoo book. It's lovely to see the herd members in the background. (© *Dublin Zoo*)

I got to meet an aardvark at the San Diego Zoo when I was there consulting in early 2023. A very interesting species.

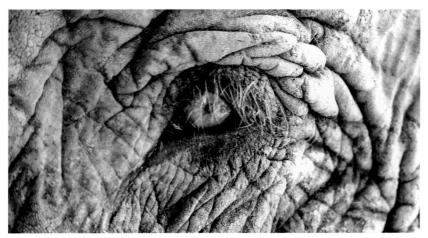

Upali, the Asian bull elephant. It's said that if you look into an elephant's eye you can see into their soul.

Showing Beyoncé and Blue Ivy around Dublin Zoo. Beyoncé was very keen and asked lots of questions about the zoo. One of the great perks of the job. (© *Patrick Bolger/Dublin Zoo*)

I was proud to lead the elephant team for so many years – a great group of dedicated and professional people behind me. The keepers give their lives to the zoo. (© *Fran Veale*)

Reflecting on the beauty of the Kaziranga landscape. A dynamic place and home to the Indian elephant in the Kaziranga National Park. (© *John Higgins*)

Looking over the tea plantations in Kaziranga National Park as we waited for a herd of elephants to pass through. (© *John Higgins*)

It was a great honour to be invited with my family up to the Áras to meet President Michael D. Higgins. His son, John Higgins, was the producer of the Dublin Zoo TV series, so he is very aware of the great conservation efforts.

The two zoos worked closely together on crate training Upali to travel to Dublin, and Alan Roocroft, keeper Ciaran McMahon and I went over to get him. Alan Littlehales, one of Europe's most experienced elephant care specialists, had worked with Upali since he was a calf, and he travelled back with us to help with the introduction process. Having a calming, familiar voice that Upali knew and trusted was key to getting him settled in the first weeks in his new home.

Cortisol checks, to determine Upali's stress levels, were carried out before he went into the crate, when he was in the crate and when he came out. However, the only time he showed any elevation in cortisol level was when he was finally introduced to the females in Dublin. From the start, he was a reassuring presence for the cows, and he thrived around them.

Bull elephants can be very dangerous, particularly when they come into a condition known as musth (a Hindu word for 'intoxicated') when they have a hormone surge and can become aggressive. Upali, however, remained steady, though he was a great indicator of when females were coming into oestrus, because he would start secreting a very musty-smelling oil from his temporal gland, in the mid-section of the head. This indicates a surge in testosterone, and we would know that a cow was coming into oestrus. It could even be a couple of the cows because elephants can have synchronised oestrous cycles. This is nature's way of ensuring greater cohesion in the group since calves born together can support one another and learn, grow and develop together. Upali fathered seven calves, and then, at the request of the international breeding programme coordinator for Asian elephants, we had to keep him separate to prevent more pregnancies. Then, just like in the wild, he moved to a different herd.

When the EAZA's breeding programme recommended Upali's transfer to Le Pal Zoo in France, I went out to see where he would be living, and to tell the keepers there about his personality and habits. A member of their animal care team also spent some time in Dublin with Upali to get to know him, and there was an exchange of dung between the two zoos (a sort of faecal text message to introduce the elephants to each other's scent and chemistry).

Health checks before he travelled included a bronchial lavage to check for TB. Despite training him for months, it was all hands on deck, in addition to John Bainbridge, the zoo's veterinary surgeon, and anaesthesia specialists from Germany. Upali was given a standing sedation to make him relaxed; this posed a risk that he might fall or hit his head on the protected-contact wall bars, so there was a team on standby to cut him loose quickly if needed. With his eyes covered, fluid was passed up through his trunk and up into his lungs, and a sample was extracted. An endoscopic camera allowed the specialists to see inside his trunk.

We used Upali's favourite foods (melons and bananas) to train him for the journey by road and ferry to France – offering them as rewards for walking into the crate and spending time there. By the time moving day came around, Upali was psychologically and physically prepared. The team were at the zoo before 5.30 a.m., even before the peacocks were up, to prepare him and to say our goodbyes, feeding him bananas for the final time. Only keeper Donal Lynch, who had worked closely with Upali, would accompany him to France to get him settled in. Once again, we were all in awe of this wonderful animal's patience, of how calmly he accepted the restraint bracelet that would keep him safe during travel and entered the crate. He didn't even object as his crate was winched onto a truck. He emerged equally unstressed on the other side.

We knew that it was time for Upali to move on to a new herd and new mates, but we were devastated to see him go. He had been such a pleasure to work with and had a hugely positive influence on the herd and the calves. He was a wonderful animal. He was very successful in Chester and Dublin, and I'm delighted to say he has become a father again in Le Pal.

THIRTEEN

ALL IN A DAY'S WORK

Being a zookeeper was never a nine-to-five job. A zookeeper doesn't switch off or go home at five o'clock. Keepers are prepared to put their lives on hold, miss family events, come in on Christmas Day, get up at two or three or four o'clock on winter nights to check on an animal, or make sure they have enough food. I could be called in at any hour or would come back to check on sick animals or newborns. My whole life was consumed by the zoo.

I have always found it difficult to take a break from the zoo, which has been hard on my wife Leona in particular. And I was always dragging my family to other zoos and animal parks on holidays. I did try to detach myself but usually failed. Even before camera phones, texts and WhatsApp groups, I don't ever remember not being in touch. We'd go on holiday, but I would check in with keepers or look in on animals. I always wanted to know how animals were doing, particularly if one was sick or if a birth was imminent. By the time I was operations manager, I was responsible for about 60 staff, and I would always insist that they get in touch if they needed me. I never found it a burden to reply to a text or take a look at a picture of an animal that was unwell. When you're as passionate about something as I am, you never quite switch off. I always wanted to keep my finger on the pulse.

I wouldn't have swapped my job for any other in the world. I loved the variety each day brought: I might need to assist with giving a gorilla a medical check or help during root-canal surgery for a tiger; I might paddle the dinghy around the lake to get a closer look at the red ruffed lemurs or gibbons or find myself wrestling a crocodile so its jaws could be secured with tape for a veterinary inspection. That last job would take three members of staff, under the direction of reptile keeper Garth de Jong, to control the croc's thrashing. A crocodile's reflexes and power are phenomenal, and they have the potential to inflict horrific injury – or even death.

For years I was also in charge of the emergency response gun crew. The zoo has a fully active firearm team, and all around the zoo are strategic locations where firearms can be accessed in the event of an emergency. We were trained by Alan Kearney, a captain in the Irish Army, who would take us out three or four times a year. It was rigorous training, on a par with army and police training, and you would have to be signed off as a qualified marksman. Training included mock situations where an animal or person might appear around a building, and you would have to decide in a split second whether to shoot or not.

A firearm team is essential to get your zoo licence. Designated members would have the firearms key and be on call every day. They wouldn't even leave for lunch. We prepare and plan for mass escapes, evacuation plans and even potential terrorist acts. Thankfully there has never been anything like that, but you have to be prepared for every eventuality. The team will even be on standby when an animal is heavily sedated and being moved for surgery. You always have to have a forward plan. We sit together and put a security plan in place.

Zoos categorise their animals by danger. Guns are only taken out for Category One animals, which are seen as an immediate threat to human life in the event of an emergency or escape. In this category are the big cats, rhinos and large apes, including chimps, but also quiet animals like zebras. A zebra is a flight animal that runs through fear. When zebras are separated from their herd or panicked, they run blind, without evaluating where they are going. In the African savanna, they would be running into open spaces, but in a zoo environment, they could run through a fence, a window or people. You would never get a chance to immobilise them, so you would have to neutralise them.

Right up until the time I left the zoo, I was also the marksman of choice when an animal needed to be darted. As operations manager, I liked that. The keepers were caring for the animals daily, so I'd say, 'Let me be the bad guy because you're going to have to go in and face the animal tomorrow.' We darted when an animal required sedation for medical care or to round up animals that had escaped. And there were escapes!

In 2008, Maggie the orangutan swung across the moat of the old habitat. The alarm was raised by a school tour, and she was found between the back of her house and the boundary fence. I was on my way to a funeral in Cabra – a friend's father had died – and my radio went, 'Gerry Jnr, over. Come back to the zoo straightaway. Maggie's after getting out.'

I went back, got the dart gun loaded and went to find her. Luckily, she hadn't tried to climb over the fence. She looked at me and looked at the dart gun, and you could all but hear her say, 'Don't do this.' She towered up and turned around, and I was able to dart her into the muscle on her shoulder; when the sedation took effect, we were able to get her back into her habitat. The next day I went up to see her. In those days, the rules on feeding were different, and you were able to give them a treat. I brought her a little bar of chocolate and a couple of yoghurts to show her how sorry I was. I sat there with her for an hour or two trying to make friends again – but it took a while!

Another time, three colobus monkeys (a very beautiful and agile African species) jumped off their island and went out into the Phoenix Park, where they were found up a tree. We had to bring in a Height for Hire machine, so I could get close enough to the tree to dart them.

We have islands on the zoo lake, which are brilliant for the animals because they can interact with the water and different species. But we didn't get all the designs right at first. We had Sulawesi-crested macaques, which are very rare, endangered ground-dwelling primates from Indonesia. I came down early one winter morning with my colleague Ciaran McMahon because there had been a heavy frost overnight. The lake had frozen, and we knew they would try to go wandering and exploring. When we got down to the lake, they were running the length of the island, sliding across the icy lake on their bellies like kids, and then running back to do it again. We had to quickly get out and break up the ice.

Another time they did get across the ice. We got a call to say there were monkeys crossing the road over to the cricket grounds. I went out in the car to have a look. I had darted them on previous occasions for veterinary purposes, so when they saw me, they set off the alarm call: 'Ooh ooh ooh, here's Creighton!' But we managed to confine them to a tree, dart them and return them to their island. They were harmless – they just wanted to get out and have a look around and a bit of a walkabout. There are now water pumps so the water can't freeze.

This was soon after we had opened our islands. It was a huge improvement on the old Monkey House, where they were all kept on sterile floors, and they were excited with their new habitats. They were in the middle of a park with trees everywhere, so it must have been incredibly tempting to get out and explore somewhere more exciting. However, we were able to make their islands more secure and give them more things to keep them happy and active, so they didn't have the urge to stray. There was never any danger with the monkeys, but they certainly kept me fit running after them or climbing trees to get them down.

I've also had to go in pursuit of birds. Before we stopped pinioning the flamingos, one of them was pinioned on both wings in error. So it figured out it actually could fly, even if its wings were very short. We got a report one day of a flamingo down at Island Bridge on the Liffey and thought it was a wind-up, but when the keepers did a count, there was one gone. I spent about 10 days canoeing up and down the Liffey trying to catch it, as it spent its day flying between Heuston Station and the Anglers' Rest pub. I was exhausted from canoeing after him and then a guy from one of the rowing clubs offered to take me in his motor-boat. We were hours following him, but eventually, he went into a corner, and I had to dive into the water to grab him – very gently, because they're very fragile birds. Despite all its flying, when we got it back to the zoo, it had put on loads of weight from all the plankton and good nutrition in the Liffey.

There's an urban myth in Dublin about a mum walking into the bathroom after a visit to the zoo to find a penguin that her kids have smuggled out. But in 2010, a penguin was stolen for real. Two young fellas were spotted climbing over the gate. They had put the penguin in a training bag and stopped a taxi. The taxi driver rang the park and told us where he had dropped them; then a pet shop owner on Parnell Street called me to say that a young fella had been in the shop trying to sell a penguin he had in his bath. I said, 'You better tell him that we're going to be coming down now to find that penguin!' We got her back after a bit of a ramble around the city. They didn't want to be caught with it, so they let it out and we picked it up on the street. Unfortunately, the penguin lost an egg at the time. She was on an egg, but because she had been taken off it for a few hours, she wouldn't go back on.

We lost penguins in other ways too. Over my 36 years in the zoo, I probably spent the equivalent of a couple of years trying to catch foxes that had broken in from the Phoenix Park. When the flamingos couldn't fly away, they and the penguins made easy pickings. There is a fence now, but foxes still occasionally get in. They're so smart. We were wondering how one fox was managing it until we realised he was following the security van when it came in during the night – the driver opened the back gate with a fob.

He would then follow the van back out later when it left!

Liam Reid, Ciarán McMahon and I could spend weeks tracking down foxes. As we got more experienced, and the zoo was getting too built up to find them, we would stake meat to the ground and put sand around it. If the fox was coming to that area, we'd see its footprints in the morning. We would feed him, creating a dependency for a week or two, so we could catch them and release them back out into the park. When you're a zookeeper, you don't want to kill any animals, but we have to have a management plan to protect our species, and sometimes we had to shoot foxes, including one that came in and killed five or six penguins in one night.

Some days, it's the humans that are the main attraction. Because of my passion for the zoo, and my knowledge, I was often asked to bring around visiting dignitaries, politicians and other famous faces. Some of the celebrities I've shown around include Woody Allen and Jennifer Lopez. J.Lo had her mum, her kids and her boyfriend at the time with her. Her mother was a great character, and she was telling me about growing up in New York with her three Irish friends. J.Lo was so down to earth and full of chat. She wanted to know all about the animals and was really interested in what we were doing in the zoo.

A couple of years later, when Beyoncé was in town, she asked me to take her and her daughter Blue Ivy around. She had a big entourage, but she was just lovely, having fun with her little girl, singing and playing, in between asking me lots of questions about the animals and the zoo's history. She also gave us tickets for her concert, and Mia, who was about eight or nine, came down to say hello.

I've always been very proud to tell visitors just how much

our zoo, and our job, has changed. Modern keepers are really architects of the animals' environment, and much of their day is spent finding ways to enrich the lives of the animals in their care, to improve their physical and mental health. Animals need to be challenged. We can enhance their lives using natural objects: a root ball full of clay, insects, trees and branches.

As we learned more about how to keep animals active, we started introducing different scents and smells. Sometimes I might even spray a can of Lynx around the big cats' habitats, and they would come out and mark their territory. Even the smell of another animal can be used for stimulation. If I did foot care on an elephant, I would bring the big pile of hard skin up to the African hunting dogs. They would go crazy rolling around in it. We have brought up a bit of tiger bedding to the elephants, and as soon as they smelled it, they all came together, closing ranks. They became protective because they interpreted this strange smell as belonging to a predator. Doing this correctly can cause cohesion and cooperation within the group without causing undue stress. In some zoos, the silhouette of a hawk is flown over the area where small monkeys live. One will let out the alarm call, and they'll all run into a sheltered area. When the shadow is gone, they will come out. It's not distress. It's a little bit of healthy stress that can bring the family together.

We're always looking for novel ways to enrich the lives of the animals in our care and give them new experiences, but only in ways that mimic or show natural hunting or social behaviour, and always for the benefit of the animal. For the smaller monkeys, like tamarins and marmosets, we would drill holes in the trees and put mealworms in them; then, they would try to grab them as they emerged. Many small primates are gum feeders (meaning they bite into the bark and suck the gum out from the trees), so we would create little cracks in the trees and put syrup in there; they would then spend hours searching for it. There are so many different things that we can do to keep them enriched. For the chimps, we might even bring in novelty items. We make up parcels with hessian sack and tie them tight; they're like kids at Christmas trying to open them to find out what's inside.

Reptiles were once put in a sterile area and forgotten. They tend to be pretty chilled, but like every other creature they need stimulation. Newer habitats provide height zones, water and branches to wrap themselves around. Different substrates within their area – grass, bark mulch, sand, hot rock, cold rock – give choice. Zones with different temperatures mean they can drift between warmer and colder spots.

For the penguins, we might load fish into something that floats on top of the water, and they have to manipulate it to make a fish fall out. Floating islands might be created for the sea lions, so they have to balance on them to get some fish.

Enriching animals' lives is just one of the requirements of modern animal welfare across what are called the Five Domains: nutrition, environment, health, behaviour and mental state. Animals have rights that need to be met. A philosophy of wellness is what inspires new zoo programmes, and understanding of the empathy, care and needs that animals have. It is a world apart from how animals were treated when my father – or even I – began work as a keeper. However, I realise we were all part of an evolutionary process. Da was doing things better than they were in the 1930s and '40s. I'm doing things in a better way than he was, and we continue to improve.

Modernising habitats has made our role less important to

the animals, as it should be. Our objective is to make keepers and humans as insignificant as possible in animals' lives. If animals have multiple opportunities in their days, and nights, they are not watching the gate for the keeper to come along so something good happens. The keeper should never be the focus. A herd or a troop or pride could be together, interacting socially, hierarchies being observed and play happening; then they hear the rattle of the keeper's keys, and they run away from each other. That's wrong.

As a keeper, it was wonderful to become unimportant to the animals. I could walk by an ape habitat, and they would throw me a glance because I was a familiar face, but they no longer saw me as the supplier of food or stimulation. It was such a contrast to the old days when I was the centre of attention. I'd be walking past the old barren enclosures, and their eyes would almost be saying, 'Please feed me, or do something to relieve the boredom.' In the old days, we were their world. Anything that happened, happened around us keepers.

In my early career at Dublin Zoo, I saw inexperience, naivety and even stupidity, but never violence or abuse of animals. Everyone did their best, and most keepers were there because they loved animals. We didn't always get the techniques right, but we evolved. Now keepers are educated, and they are part of a network of zoo people across the globe that share information about different species.

To become a keeper, you need to do a zoo animal management course, so the job is a combination of education and practical experience. Education is a wonderful thing, but I do think that initials after names have become a little too important. Some of the best staff I ever had on my teams began working on the grounds picking up litter. Being able to look an animal in the eye – and have an animal respond to you – is a unique gift, and you cannot learn that in college. I do sometimes feel there is a conflict between modern approaches and solid animal relationships.

I've seen so many people without education be incredible keepers. John O'Connor, from Ballyfermot, joined the zoo at 15 and was probably one of the best giraffe keepers I've ever seen. He worked in the zoo for over 52 years, as did his father, and his knowledge and understanding were astounding. He could tell days before, just by looking at a giraffe, when a calf was going to be born. I used to see him walk among the giraffes and think, 'Oh my God, he's going to get killed or kicked,' but of course that never happened. I could never have done what he did. He had confidence because of the unique relationship and bond that they shared. The giraffes just loved him. He and keepers like Eddie O'Brien, Helen Clarke and Joe Byrne were the corner-stone of Dublin Zoo: hard-working, solid keepers who learned on their feet. They knew their animals intimately, and it was a very personal form of zookeeping. They are part of a core of people who had a historic connection with the place, not to mention love and loyalty – they gave their lives to Dublin Zoo.

Like them, I always cherished the privilege of working with incredible animals and the uniqueness of what I got to do. I would be doing a pedicure for the elephants and saying to myself: 'I'm the only person in Ireland doing this today. How wonderful is that?'

Avani

Every day as a zookeeper can bring surprises, but we were caught completely off guard one morning in March 2017, when we came in to find that Dina, our matriarch, had given birth. Generally, elephants have a four-year gap between births. They start pushing the other calf away, and another one comes along. However, Dina's other calf, Samiya, was only about two and a half years old when she gave birth again.

Dina must have come into oestrus almost immediately after giving birth to Samiya – normally that wouldn't happen when an elephant is in human care. In harmonious herds in the wild, where resources are very good, interbirth periods can be shorter, and other elephants alloparent – meaning they take over suckling and caring for the newer calf. Alloparenting has been observed in elephant society when a mother is sick or dies, or if a calf is rejected; then, a grandmother or related elephant will help supplementary feed the calf. It's not unusual for elephants to show that incredible sensitivity and caring behaviour; but this happening in Dublin Zoo was a testament to the harmonious situation we had created.

We don't know what communication goes on within the herd, but when Dina decided she wasn't feeding another calf, her older daughter Asha immediately took over.

However, Asha had her own calf to feed. Zinda had been born the previous September, and her nutritional demands were very different. Newborn milk would include colostrum and have a very high fat content, so although Avani was feeding from Asha, she wasn't getting anywhere near the nutrition that she needed, and she deteriorated before our eyes.

In addition, Asha would allow Avani to feed only when Zinda was suckling, her instincts and loyalty being towards her own calf. She would allow the two together, but if a hungry Avani tried to feed ahead of Zinda, Asha would kick her away. We could see the stress being caused within the herd.

The demands were too great on Asha, who wasn't able to support two calves. We wanted to leave Avani in the herd and tried everything to get her to take supplementary feeds, but after about six weeks we were gravely concerned. Avani looked thin and weak, her skin was all shrivelled and dry, and we could see bones protruding. We put the vets on standby in case a decision was made to give her critical care by tube feeding her.

It was a really difficult situation for us to watch as a team. We had never had a crisis of this proportion. We thought we were going to lose her, and we were even planning how we would deal with her death. We tried everything to encourage Avani to come close to us at the protected-contact wall to offer bottles, but she didn't understand what we were offering. We were only just another object in front of her.

Then I had the idea of trying to get jets of milk close to Avani's mouth area, thinking that tasting some milk might get her interested. I tried once or twice with large syringes, but she was pulling away because she didn't know what was happening.

The elephant houses have an irrigation system that we use to shower the animals. So we turned on the showers. Not only did it hydrate Avani, but it also acted like warm rainwater on her skin, and we witnessed some play behaviour as she tried to move her trunk and interact. Then, when she was desensitised to the touch of fluid, I filled a kids' Super Soaker water gun with milk and started to squirt it just below her eye line, in a position at the top of her trunk where it would roll into her mouth. We could see she was picking up the taste and starting to lick it, and I also squirted it around her feet so she was getting this milky smell and milky feel. Almost immediately, we saw a positive reaction. We were pairing that with voice calls, saying, 'Good girl, good girl', and every time we called, we would squirt the milk, so she would have a positive association with the human voice and the milk hitting her face.

Avani started to gravitate towards the sound of voices because that sound meant milk was coming, and eventually, she got really close to the protected-contact wall. Soon she was so comfortable coming over to the wall that we were able to open one of the foot ports for the adult elephants and squirt some of the milk from a 50ml syringe into her mouth. She took it in seconds!

It was a joyous moment. The Elephant House was awash with celebration and happiness. Keepers were crying, and I was on the verge of tears myself. It was an incredibly emotional moment to know that we had maybe finally turned a corner with this little fighter, this little survivor who just wouldn't give up on life.

It was an amazing feeling to get that first bottle into her. After that, the elephant team, which is always brilliant, worked around the clock to feed her within her family and keep her under observation. Avani would come through the calf door in the protected-contact wall for her bottles, and then she'd return to her mother and the herd. We would place her trunk back on the head during feeds to mimic the behaviour of when she was suckling, the natural, comfortable position. She thrived and excelled, and it wasn't long before she was drinking 25 litres a day!

It was so good to look at her early in the morning on the cameras and see her fast asleep on the ground, and not bawling looking for a feed. Every day we could see her progress. We celebrated as the wrinkles and fat started to appear on her back. Her weight was monitored daily, and when she started putting on a kilo a day, we knew we had turned a corner.

Our little Avani, our little fighter, would grow up!

FOURTEEN

ON TV AND IN THE WILD

In the 1980s, Da was featured on *Bosco*, when Bosco would go through 'the magic door' to visit the zoo. He also appeared on *The Late Late Show* and other TV series from the zoo hosted by Peter Wilson.

As I grew older, I was also always happy to talk about animals in front of crowds and cameras. One of my earliest appearances was in a documentary called *Keepers of Wild* alongside Peter and colleagues including Eddie O'Brien, Joe Byrne and Peter Phillips.

I became a kind of poster boy for the zoo, going out to give talks and make television appearances. I'd be on RTÉ Television weekly, on *The Den*, talking about animals with puppets Zig and Zag and their human presenters, including Ian Dempsey, Simon

Young, Ray D'Arcy and Geri Maye. It was great fun. The first time I went out, I had notes about what I was going to say, but I quickly learned to throw any notes away, to expect the unexpected and go with the flow. It was brilliant craic.

Everybody wanted to know about the zoo again, which was great, and people were saying, 'Isn't it great to see the changes?' Those changes coincided with the filming and broadcast of *The Zoo* TV programme, a series that became instrumental in changing how the public saw us.

It was 2009 when Shane Brennan and his team from Moondance Productions first brought the cameras to us to follow the joys, trials and tribulations of zoo life. Director Leo Oosterweghel only allowed the cameras in when he felt that the zoo's infrastructure was up to international standards, and when the philosophy that puts animal wellness at the centre of everything was ingrained in the staff. Before we began filming, Leo told me that he wanted the programme to be a true representation of zoo life.

People describe the show as an emotional roller coaster. There are births, deaths, serious surgical operations and animals arriving and leaving. We showed a macaque monkey carrying her dead baby around for two days, and letting apes come in and witness a dead group member. We had a wonderful team of both old and young keepers who all spoke from the heart. I loved being part of it, telling the stories. People loved the show because they got to experience our joy and our pain — because that's what working with animals brings you.

The Zoo changed our lives as keepers, and it changed our status, making us proud of who we were and what we did. It also contributed to the huge cultural shift in the people who visited:

the zoo was no longer a dumping ground for old fruit and bread, and it was rare to see anybody tease or shout at the animals. Visitors now knew they had to be respectful. This was a reflection of the ethos and philosophy instilled by Leo that animals come first; the needs of the animals are not necessarily what the visitors want. People began to realise that if they didn't see an animal, they needed to be patient, that they were viewing these creatures on their terms, in their territories. There were beautiful areas for people to view the animals, but the animals were empowered to manage and control their days.

The programme also educated people about nature and the roles of different species. For instance, we love reptiles in our house – we keep snakes and geckos – but a lot of people say they don't like them. However, learning about their ecology and biology and the role they play made many people view them differently.

I'm not a celebrity, but it's wonderful to be stopped walking down the street, in the shops or at the airport by people wanting to ask 'yer man from the zoo' about a particular animal. *The Zoo* was shown all over the world, and I heard once that it reached an audience of over 150 million people. Tourists would come who had seen us on TV and would stop us to talk. This little zoo in Dublin, Ireland, was having an international impact.

I was walking around the zoo earlier this year with Tommy Tiernan, his wife Yvonne and their children, and because we were on my home turf, more people were stopping me than him. He asked me if I ever got tired of people asking me questions. The answer was never. The average conversation I'd have with a person would be about 50 seconds, and they usually just want to ask me about a particular animal they have seen on the TV show. I'm happy to do that all day long if it influences a kid to learn, or if it spreads a good message about conservation.

One of the real tests for me during production of the show was a decision to show the last moments of the zoo's old lioness, Sheila. At 25 (a phenomenal age for a lion), she was the last remaining member of a long and distinguished family of African lions who had lived at Dublin Zoo. I had looked after Sheila from a young age. I remember vividly coming in on New Year's Eve in 1987 to see four beautiful tiny cubs snuggling up to their mother. One of those was Millie, the Millennium lion cub I hand-reared, and another was Sheila. Now Sheila had to be euthanised.

That decision wasn't taken lightly: the policy at the zoo is to allow animals to live out their natural life. It is only when their well-being and quality of life are compromised that difficult choices are made, and the team unanimously agreed that it was the right choice for Sheila. We had noticed a sharp deterioration in her health, and she wasn't even willing to stand up. As she fell asleep and took her last breath, she was watched over by the vet, her keepers and lots of other staff who knew and loved her. Everyone gathered to pay their respects, and it was all recorded and shown on the TV show.

We were worried about the public backlash. Thousands of emails arrived the next day, but all of them were asking if the keepers were okay. For me, it was a realisation that people finally understood what we are about. Nobody questioned our decision, because they knew that we cared for and loved the animals.

In 2019, one of Moondance's directors/producers, John Higgins (son of President Higgins), accompanied me on a visit to Kaziranga National Park, the wildlife sanctuary in the state of Assam, northeastern India, that had been the inspiration for our elephant habitat. Dublin Zoo is very involved in wild elephant conservation, in particular supporting the development of elephant corridors, one of the most important solutions to stopping population decline in their native habitat.

Kaziranga is on the southern floodplains of the Brahmaputra River, which originates in China and flows through India and Bangladesh. It is regarded as the only 'male' river in India because of the strength of its waters. Every year, these waters erode masses of soil from its banks, depositing it elsewhere. During the rainy season, it floods most of the park. The Brahmaputra is both a giver and a taker of lives. The water rejuvenates the plains and the plants, bringing life, but the floods kill hundreds of animals, including elephants and tigers, while others are forced to search for safety on higher ground.

Even outside the floods, elephants and other animals often leave the reserve in search of food, but the busy national road going through the park makes this very dangerous for both animals and humans. We were guided around by wildlife biologist Seema Lokhandwala, of the Asian Nature Conservation Foundation, which is involved in the conservation of Asian elephants in South and Southeast Asia. One night, we joined rangers as they stopped traffic on the road through the park to allow the elephants to cross, on a route they have traversed for generations. The herd, including a calf of less than a year, came climbing up a steep bank and crossed to feed on fresh bamboo on the other side. They would cross back hours later. Other animals, who also use the corridor, were also waiting in the darkness: we could hear sounds, and the rangers were keeping a close eye on some water buffalo. They looked quite docile, but the bulls are responsible for more human deaths every year than any other animal in the park.

The elephants are tracked throughout the forest, so armed rangers can predict when they're heading for the road. If they don't stop the traffic and manage it, everyone is jumping out to take selfies or beeping horns and trying to chase the elephants, and there have been fatalities on both sides. Being there brought home the reality of the race for space and the human/elephant conflict that is happening across the Indian continent and in other countries. While we were there, seven elephants were killed on railroads. One calf was killed, and the rest of the herd were circling its body when another train came. It's just one example of how the human domination of a landscape impacts wildlife.

This is an area where humans and elephants coexist together – not always successfully. Wild elephants are being pushed into smaller and smaller pockets. When they leave the park, it can have disastrous consequences for surrounding fields and villages, as crops can be damaged and irrigation systems are torn up as elephants look for water. I went to visit one village where they dread the arrival of the elephants and the damage they may cause. Yet, they realise how special these creatures are and consider them good luck.

The first time we were out in the park, a majestic bull elephant came out of the forest and was only five or six metres away from our jeep. He was magnificent. His ears were flapping, probably to cool his body, but elephants also flap their ears when they are angry. Careful not to alarm or annoy him, we reversed out of his way. Then he just vanished into the high grass again. I was in awe watching how he operated, and how quickly he could immerse himself in the landscape and disappear.

It was just one of many magical moments for me. That same day we heard an alarm call being sent out by a barking deer. One of the rangers told us there was a tiger on the move, and as we came around the corner, there he was, sitting in the grass. He was breathtaking. We saw wild boar, swamp deer, hornbills (just like the ones in Dublin) and different species of monkeys. We saw quite a lot of buffalo, shadowed at all times by little egrets, which love to eat the insects and small creatures that are revealed as the buffalo move through the swampy waters, churning up the soil.

The egrets also love the great one-horned rhinoceros, and seeing these magnificent, endangered beasts grazing and moving through the park was one of my trip highlights. They are very different from the African rhinoceros. They have large, armoured skin flaps, or folds, on their body (which help regulate body temperature) and large incisor teeth, which they use for fighting, rather than their horns. At one stage, there were only about 12 individuals in Kaziranga; now, there are over 2,000, which is wonderful. We came across a mother and baby, and then another two grazing in the marshland area. The big male was probably around 2,000kg in weight and behind him was a troop of maybe 70 or 80 rhesus macaques. To see them in their natural habitat was a dream come true.

I also encountered a muckna, a bull elephant born without tusks. Being born with tusks dramatically decreases an elephant's life expectancy, because of ivory poaching. As a result, a lot of African and Asian elephants are being born without them – nature is trying to save them. Mucknas also tend to have thicker, stronger trunks than tuskers.

On my last evening, we got to watch a herd of nine elephants crossing the river, the younger ones having to swim. After working with elephants for three decades, to see this multigenerational herd interacting in the wild was incredibly emotional. Even though I was being filmed, I found myself so in awe that I couldn't find the words I wanted.

Kaziranga turned out to be far more than I had ever expected: the diversity, the abundance of wildlife and the conservation work were everything that a zookeeper like me would want to witness. It was just remarkable, and I was inspired with more ideas for our herd in Dublin. It was also inspiring to see how conservation is working in India, and how animals are being protected, though I was less comfortable and more pessimistic after meeting some modern mahouts.

Mahouts are the men, and boys, who train and work with India's elephants. It is a tradition that has been passed down from generation to generation for thousands of years. Owning an elephant is a big deal. They cost a lot to feed and look after, but they also provide a means of supporting and feeding a family – often the only means of survival.

Each elephant has two carers: one who works the elephant and another who looks after and feeds it. Watching the mahouts bathe and feed their animals, you might feel respect for their unique relationship. But from the human tone of voice, it was obvious that this relationship is based on fear and discipline. The elephant knows that if it does not comply with instructions, the mahout will jab it with a long stick with a sharp, pointed steel end. The elephant understands that there is a punishment if it doesn't comply. Essentially an animal's spirit is broken so it can be managed. I found it very upsetting to watch.

India is the country where the elephant is said to be revered, the country of Lord Ganesh. Elephants are revered and supposed to be protected, yet one cannot ignore the contradiction of mahout training or the abuse they suffer in temples, where they are kept chained and displayed. This is not about culture. Culture has turned into commerce. These elephants can earn thousands of dollars a day to be hired for the temples.

The film *Gods in Shackles* is a heartbreaking portrait of what still happens to elephants in human care. An animal that is biologically programmed to move, be social and search for food for 18 hours a day is expected to stand in a procession for a couple of hours with fireworks going off, surrounded by noise and light. They are trained with violence to accept these conditions, taught that there will be painful consequences if they move from that position. When the elephant is young, it is tethered to a tree and a long cane is placed behind its ear. If the elephant knocks it over or lets it fall, it gets beaten by the mahout – and this could happen hundreds of times in a training session. So when you see these bull elephants with a cane behind their ear, standing perfectly still, it's because they know that when that cane drops, there will be severe and painful consequences. The chains placed on an animal's leg may also have spikes pointing inwards, so if the elephant pulls, it causes pain; and sometimes the elephants are intentionally blinded in one eye (on the side where the mahout walk) to make them more manageable.

In addition to changing how we manage elephants in zoos, it is vital that we challenge the cultures that deprive them of any kind of proper life in their home countries. In order to do that, we need to support communities where the working elephant is the source of family income, to help them find an alternative way to survive and thrive.

FIFTEEN

TINDER FOR TIGERS

Everyone loves when a new baby is born at the zoo, but it's not as simple as letting nature take its course! Many species are part of international breeding programmes for endangered animals. There is a studbook database, a sort of international dating agency for animals, and studbook keepers who manage those dates, deciding who will make good breeding couples.

Dublin is part of the EAZA and has more than 35 breeding programmes. Approximately 350 zoos get together every year with all the regional studbook keepers, to plan out strategies, decide which species need help and support, and which animals, in which zoos, will be bred. It's very detailed and scientific. Ultimately, it's all about keeping the gene pool as diverse and strong as possible. Dublin currently coordinates two European breeding programmes: citron-crested cockatoo and Goeldi's monkey.

Dublin also has Amur tigers (there are less than 1,000 left in the world) and Asiatic lions (just 400 in the wild). Without these breeding programmes, these would become extinct species. The studbook keeper knows all the Amur tigers in the European population, and they might recommend that the Dublin male goes to meet the German female – a sort of Tinder for tigers.

Nowadays decisions to move animals from their social groups to other zoos are not taken lightly. Animals like lions are very sociable, but males are biologically programmed to drift in and out of prides, so there are fewer consequences with moving cats. We will notice animals showing signs of grief, but they get back to normality very quickly because the integrity of the herd or the pack or the troop is what's important. However, when we know that an animal is moving, we may start the process months in advance, so that every animal in the pride, pack or troop comes to an understanding that a detachment is taking place. The separation needs to happen over time, so they are prepared for it – psychologically, physically and emotionally. In the past, we didn't understand the implications. If there was a troop of 20 macaques, the studbook keeper might say that five of them need to go to another troop. Now we know we cannot just whip the five out one day and expect there to be no consequences. There needs to be a gradual separation, maybe starting by moving them to an adjoining island or habitat where there are still vocalisations and the animals can still see one another.

However, even endangered species cannot be continually bred in a zoo situation, because we have to maintain genetic diversity. We have to make sure that all the animals born can be housed correctly and remain part of the European Endangered Species Programme (EEP) breeding programmes. We don't want them to get into the wrong hands, so we don't breed animals if there aren't properly run zoos to take the offspring. There's always a population management plan, so one zoo will be allowed to breed this year, and another will be told to put its female on contraception for a year or two. While it might seem unkind to stop animals from having babies, it's not: contraception is probably one of the best things that has happened in managed breeding programmes.

There is a deliberate strategy among zoos, a collective effort to breed endangered species at a rate that the programme can manage. We cannot have breeding for breeding's sake, because the welfare of the animals would be compromised. No zoo has the facilities to keep breeding tigers, no matter how endangered they are. I have seen tigers becoming extremely aggressive towards their cubs at 18–24 months old because they feel it is time for the cubs to go, and breed again. This is what would happen in the wild. However, in the wild, these cubs can move away; in a zoo, they can't, so they have to be put into a backyard or somewhere where there may be compromised welfare. In the past, animals in Dublin Zoo could be kept in small holding cages for months on end waiting to be offloaded. Contraception, population management and long-term breeding strategies are the way forward.

There are Sulawesi-crested macaques up at the zoo. In their home country of Indonesia, they are considered a delicacy – their skulls would be bashed open, and they would be eaten at weddings. Their population was down to about 100 individuals, but because of successful zoo-based breeding, there is now a sustainable, viable population with reintroduction programmes starting. Now, to keep family groups intact, males can be vasectomised, and females given contraceptive implants.

We were also breeding wolves. They're not an endangered species, and they could have 10 or 12 wolves in a litter. So we decided that the best thing to do was vasectomise the male. He was not castrated, and he could still mate every year, maintaining normal wolf behaviour.

Primates might get a hormonal implant or a coil. Wendy the chimp would be given her pill every morning in a piece of banana. I was chatting to these real Dublin women one day in front of the chimpanzees, and they asked why the chimps were not having babies. I told them that all the females were on the pill. 'But,' they asked, 'why do you have them on the pill?' I pointed to Austin, the male chimp. 'Because it's much easier than trying to put a condom on him.' They got a great laugh out of it.

The international studbook keepers stopped Dublin's chimps breeding because, when a genetic profile was taken, it was claimed they were hybrids, not a pure generation of chimps. Yet we know they came from the wild – some were even sent here by an Irish missionary priest in the 1960s. And then they did something I always questioned. They decided to bring West African chimpanzees, a pure species, into the breeding programme. All these chimps had been taken from the wild and brought to the Netherlands for AIDS research. They were kept on their own, so they were socially bankrupt. They were then put into zoos – including Dublin – and they didn't know how to behave or how to mate. Chimpanzee societies don't function without babies in them, so they broke down all across Europe because of this stupid idea of a purebred chimpanzee. And now they're telling people to go back to the original breeder chimps and manage themselves.

Even the decision to breed endangered elephants, as we did in Dublin, has to be considered carefully. If a herd produces just bull elephants, what is going to happen in five years' time, when they start to get pushed out or pushed away? Where are they to go? There is already an issue of 50 surplus bull elephants in Europe because breeding has been so successful.

One of the consequences of animals being in human care is that we tend to overcompensate with a really good diet, and this has been linked to a rise in the number of male births. The EEP is now recommending all-male groups of elephants, as there would be in the wild, and some zoos and parks keep bachelor groups of elephants and gorillas (male gorillas born in Dublin are sent to Longleat safari park, for example). They live happily until you bring in a female – and then it all goes out the window!

When breeding – whether arranged or impromptu – is successful, everyone gets very excited about the idea of a new baby. When we were waiting for our elephant births in Dublin, I was like an expectant father. It was a hugely emotional event for all of us. It's hard to describe the relief and excitement of the whole team when those first calves were on their way. I didn't sleep well at all. I'd wake at 1 a.m., check the cameras and then sleep for another couple of hours.

Over a decade, I was privileged to share – from afar, via cameras – nine elephant births in Dublin's herd. It is a most amazing sight. The celebration of a new life coming into a herd is one of the most exciting things I've ever seen. Everybody's involved, flooding the calf with a massive vocalisation of sounds to welcome them into the world. A new birth is always greeted with excitement, ears flapping, tails slapping. Young female elephants are pushed in to learn, to see the birth process, to witness and hear the sights, smells and sounds that are unique to this moment. Young bulls are pushed away because they have nothing to offer. And as new life stumbles onto the sand floor, there's this absolute orchestra of squeaks and thrills and excitement. The older cows kick and push the newborn to stimulate it to stand up – it looks quite violent but it's not – and there is inch-perfect manoeuvring around the newborn as the herd creates a sort of mayhem to protect it and distract any predator that may be lurking nearby. It is a masterclass of multigenerational cooperation and learning.

The herd stays around minding the calf, as they would in the wild. Everyone has a role – from encouraging it to stand and suckle, to trampling on the afterbirth. The mum will touch around the calf's rectum, which stimulates it to pass urine and faeces for the first time. This is all normal, to make sure that the calf is healthy, vibrant and ready to move along with the herd. The afterbirth is kicked, passed around and torn up, to mask the smell and distract predators, just as they would do in the wild. They let the calf know that it's safe, protected and part of a family.

During its first couple of hours of life, you want to see that the calf is moving well and that it has figured out the technique of folding its trunk back on its head to latch on and suckle. Suckling produces colostrum, which is vital for a calf's immune system.

Herd behavioural changes coming up to a birth are significant. The dead giveaway that something is happening is closeness. Most of our births took place at night, but when Yasmin gave birth to Kabir it was during the day, and we got to see everything. The younger females were just stuck to her like glue. They became a little closed group, touching her gently, putting their trunks up and smelling around her genitals, offering tactile support.

Sometimes we'd see the younger elephants having a particular interest in something on the ground around the pregnant female. If it was a large mucus plug, then we knew it was game on! The body language of an elephant in labour can be quite dramatic. She can throw excessive sand, or urinate on the sand and throw it over her body, as if trying to cover her body with its scent. Yasmin's ears were flapping, and she would raise her tail and bang it down towards the birth canal. She was also stretching and bending down on the back legs to relieve pressure off the cervix. Finally, we could see a big swelling underneath the tail, and the calf was pushed over the high hip bone and into the birth canal before being flushed out quickly.

When Asha gave birth in 2016, it was an unusual birth. Generally, elephants are born with their back feet first, but Zinda was born trunk first, followed by her head and front legs. It looked really awkward, but she eventually got out. Normally the mother kicks the calf to stimulate it, but with Zinda, it was just the most gentle prodding and touching. In previous births, the calves were all standing within 10 minutes, but this newborn lay down for over 30 minutes, perhaps because it had been a long labour, and the herd wanted to give the calf time to recover. However, Asha retained the placenta, which stopped her from producing milk and made her very sore and tender. She was attentive and showed all the right maternal instincts, but she wasn't allowing this beautiful little calf to suckle.

It was the first blip we'd had in 10 years of the elephant programme. We had to decide what to do: whether to allow her to die in the herd or take her away. What about the stress to the herd if they could hear her squealing in another part of the zoo? I decided that we needed to supplementary feed that calf while keeping her in the herd. It was a fine balance of keeping her alive but not too dependent on us. We were trying to keep her nutritionally satisfied but still hungry enough to pester Asha to be fed.

In the old days, we would have pulled Zinda out immediately and hand-reared her, without thinking about her future life. Now we consider the consequences of our actions. It was a very proud moment for us all when we stopped feeding her after a few weeks, to see how she was interacting, socially and behaviourally. Zinda was a fully participating member of the herd, she had her role, and she now had a future as a breeding female. I looked at her and thought, 'We did this right.'

We have trained elephants with kindness and positivity to accept human care. But humans cannot teach elephants anything about being elephants. Elephants learn from other elephants, and Dublin's success is based on the fact that it is a multigenerational herd of elephants teaching elephants, with a matriarch and her daughter at its core. Mothers encourage younger females to play and interact with their calves.

I've seen two- or three-year-old females in Dublin who copy the big females when a calf is born, lifting their leg forward to try and suckle it. That's because they've seen the mothers do this. They are not anywhere near mature, but already they have these learned behaviours.

Female elephants stay with their mothers for life, but in the animal kingdom, one of the longest relationships between a mother and their offspring is the orangutan. They will keep their offspring with them for maybe nine to ten years, during which they need to teach them everything. Chimps and gorillas have family units that support each other – chimps have even been observed minding members of the group with disabilities because their structure is so organised – but orangutans are solitary. By the time they leave their mum at 10 years of age, they have to know what trees to eat from, what predators are, what fruits are the nicest, and what pathways to travel.

In the past, young animals were often sent to other zoos before their time. Not only did this compromise their development, but it must also have caused great distress and trauma to both young animals and their mothers. Thankfully, good zoos have learned the importance of keeping animals with their mothers for as long as would be normal in the wild. Zoos need to follow natural life cycles. It is imperative that animals are given the chance to gather all the information they need for their future. If animals go away before they are ready, immediately they're on the back foot. Socially they will not be competent, or ready to be introduced to another animal. Timing is critical. We need to wait until they are almost being pushed out by their mother – where a tigress is starting to growl at her cub, for example – or they are starting to separate themselves.

For me, the start of animal welfare should be an acknowledgement that animals love and are loved, that they understand birth and death. Some people believe you can't say that animals love because it is anthropomorphic. I call bull**** to that opinion. I have seen primates, elephants and even big cats demonstrate love. If love is defined as 'feelings of strong affection for a particular individual', I've seen the greetings that big cats give each other after a prolonged separation, both in the wild or in human care. I have witnessed incredible tender moments as male adult lions and tigers were introduced to their cubs for the first time after birth. So many animals show a need for tenderness, touching, joy, empathy and being together. If that is not love, then what is?

Zinda

When training keepers, one of the most important bits of information I pass on is a checklist of symptoms for elephant endotheliotropic herpesvirus (EEHV), including swelling of the head, neck or trunk, oral ulcers or a blue tinge to the tongue. EEHV manifests itself in times of stress – a bit like humans and cold sores from the herpes simplex virus – but it can be devastating to an elephant programme.

EEHV is one of the biggest killers of Asian elephants, both in the wild and in human care (African elephants are less prone to it). Therefore it's essential that anyone caring for elephants is aware of its symptoms, how to diagnose it and how it should be treated. Early detection of viral levels means early treatment and a higher chance of survival.

EEHV affects the heart and the blood vessels, causing

haemorrhagic disease in young elephants. It's a horrible, horrible virus. If severe, it can cause death within 24 hours of infection. One of the first places elephants will show signs of infection is the mouth, which is why we train calves from a very young age to allow us to check their mouth and temperature daily. We check for colour change because the virus can cause a vessel eruption in the tongue. We also take weekly blood draws from their ears. Samples to test for EEHV used to have to be sent abroad, but they are now checked at the Irish Equine Centre in Kildare.

We had suspected that Anak and Asha had EEHV when they were younger, but we were never able to confirm because they hadn't been trained to allow blood draws. However, we activated our treatment protocol, and they were fine. Our first confirmed case, diagnosed during a routine trunk wash, had been in Kavi, one of our young bulls, but he responded very well to treatment; luckily, he had been trained for rectal fluid hydration and to swallow medication in capsules.

Then, in September 2018, we noticed that playful two-year-old Zinda, Asha's daughter, was showing symptoms: she had a swelling under her neck, and her tongue was swollen and dark. We thought we were going to lose her, which would have been devastating. Chester Zoo, which has helped us greatly with our elephant programme, has now lost seven calves to the virus. It's recommended that treatment is activated when the viral load reaches 5,000 particles per millilitre, but the viral load can accelerate rapidly: you could do a test at two o'clock, and it's 4,000 viral load per millilitre, and by night-time, it could be a million.

Zinda was treated with an antiviral herpes drug called Famciclovir, which is used in humans for shingles. There are differences of opinion about whether it works or not. Some scientists say you might as well be throwing them Smarties, while others say it works at certain times with a protein in the virus that stops it from progressing.

Famciclovir is made for humans, so an elephant might need 60 to 100 tablets two or three times a day, and they don't taste too good. We train elephants to take tablets, but there is always an issue with them spitting them out. It was a real challenge to get the medicine into Zinda. We tried it in bottles of milk, but it was still too bitter for her, even with added sugar. We stuffed them into bananas and Nutella sandwiches, but nothing worked. One day, the team spent an hour trying to get her to take six tablets – she needed 18! Finally, we ground them down into powder form and painstakingly filled gelatine capsules. We had Monkey Chow biscuits, which are hollow in the middle; we packed the capsules into the biscuits and trained Zinda to take them into her mouth, followed by a squirt of juice.

The team put in an incredible effort to get her well and keep her hydrated. Every hour, pictures of her side profile (where you could see the swelling) and her tongue (to monitor the discolouration and bruising) were taken and sent to the vet.

When an animal is sick, keepers will disrupt their own family lives, staying long hours and coming in through the night to administer medicine or treatments. We kept Zinda with the herd, and it was really interesting to see how quickly the group adapted to us coming in at night. Because we are coming in to give medication, all the herd was rewarded. We would drop down hay nets or give extra nuts. Little Avani got a couple of extra bottles. So, they would be very happy and excited to see us, and when they could hear us coming down through the zoo, would start to vocalise. It was always a lovely welcome.

To our relief, Zinda pulled through and continues to thrive. Dublin Zoo has been lucky: we had a few positive cases, but no deaths. Some we would never have known except for testing – they never showed symptoms.

There are a couple of strains of EEHV, and whatever type Chester's elephants got, they were dead within two or three days. Now, in 2023, I am hopeful that we are turning a corner. Chester Zoo continues its efforts to combat this deadly virus, and last year, with scientists at the University of Surrey, it began a trial for a potentially life-saving vaccine. Also trialling a vaccine is Houston Zoo in the US, which has been at the forefront of EEHV research since 2010 with Dr Paul Ling, a human herpes expert from Baylor College of Medicine.

I am optimistic that it may soon be possible to stop this disease. And that day cannot come soon enough. Not only can EEHV devastate an elephant programme – it is even more devastating for the herd that loses an elephant because that's what elephant society is all about. It's about the calves. It's about the next generation.

SIXTEEN

DIFFICULT DAYS

When people ask what the mortality rate at the zoo is, I say 100 per cent. When animals are in human care, sad things can happen, and this is no reflection of a lack of care or kindness. There are always trials and tribulations, and the death of an animal is always difficult – for its family and its keepers. Some deaths are more difficult than others.

In 1982, the year before I joined, there was a tuberculosis scare. A male Siberian tiger grew ill and died, and the post-mortem revealed he had TB. All the Siberian tigers – his mate and their three cubs – had to be put down. As a precautionary measure, the lions that shared the same building – three adults and three cubs – also had to be euthanised. There had been no tests available to check if the lions were infected while they were alive – and it turned out they were not. Da says it was the most stressful and difficult time of his zoo life, having to load the cats into the van to bring them up to Abbotstown, where the Veterinary Research Laboratory was at the time.

One of the darkest days in my zoo career happened in April 1997, when I was still a junior keeper. It was a lovely spring day, and I was in the yard with my dad. There was a lot of activity because a rhino was coming in and another was leaving. Dorothy, a beautiful four-year-old white rhino who had been born in the zoo, was on her way to Longleat safari park in England. I turned to Da and asked if I should bring the rifle over just in case. I was part of the zoo's gun crew, trained for emergencies and escapes. And he said, 'Sure. Do that.' The gun was a very old Lee–Enfield .303 rifle that had been given to the zoo by the army in nearby McKee Barracks. It was a standard military rifle, often used as a sniper weapon, but it was still very powerful.

Dorothy was in the stalls of the old Rhino House, and the driver from Chipperfields, who was transporting her, had to reverse up to a gate that was like a passageway at its rear. He lined up the truck, opened the back and pulled down the ramp. Then, when Dorothy's stall was opened, he started roaring and shouting, trying to get her to go up the ramp. She could see the truck in front of her, but she could also see a gap up its side. There was a panel there, which was meant to guide her into the truck, but it wasn't fixed, and in her panic, Dorothy started to push through this panel and into the public area, where I was standing.

It was early afternoon and very busy. I could see she was getting out, and looking behind me I saw women with prams and kids walking around. To say it would have been a catastrophe if a white rhino got out and started running around the zoo is an understatement. Dorothy was at least a tonne and a half, and there were people everywhere.

She was coming along the side of the truck, and fellas were climbing up trees and keepers jumping into the lake to get out of the way. I knew I was going to have to shoot her.

When she got out past the side of the truck, I was about five to seven metres away, on a bit of a hill. She came charging at me, and I thought, 'I'm probably going to die here.' I suddenly realised that behind the fence opposite me was the Phoenix Park Polo Grounds; if I missed the rhino, the calibre of the bullet I was firing would go straight through the fence, and potentially kill somebody in a car or walking in the park. So I stepped down the hill a little closer to her, to ensure that if I missed her, the bullet would go into the ground and not through the fence. My thinking was, 'I won't know anything about it, I'll be dead. But I won't have killed anybody else …'

It was my boxing brain that helped me, because I was able to evaluate, and I had spatial awareness of the catastrophic consequences. My brain went into slow motion: 'Move back down this hill and take your shot when she drops her head down' (that's what attacking rhinos do). That's what would happen in the ring, too: your brain is so heightened that you can see where the next punch is coming from. Just as Dorothy was about two metres away from me, she dropped her head – either to put the horn through me or to try to get me out of the way. I discharged the rifle, and the bullet went in just above her eyes. She dropped to the ground – it was such a powerful clinical shot, she never even kicked a leg. Her life just ended in the flash of a gun. I remember looking at Da, and him looking at me, and I could see the shock on his face.

Blood was starting to come out of her nostrils; it was a horrible scene. I handed the gun to one of the guys and said, 'I need to get out of here.' I was so upset. As a zookeeper, my job was to care for animals, so it was just the worst possible emotion to know I had taken the life of one of those animals. This was an endangered white rhino, and I had killed her. Dorothy had been in the zoo as a calf, and she was loved, particularly by her keepers Matt Wilson and Shane McCrory, who were there that afternoon.

I knew I'd done the right thing – my reaction had saved lives. Yet I felt like I'd betrayed Dorothy. Animals sometimes had to be put to sleep, and that was sad, but this was an animal that had a future ahead of her as a breeding female and was going to make a contribution to saving her species. I felt I had wasted her life. I was very distressed. When I went home that evening, my mother told me that she had never seen Da so shook. My father's a tough man, but he had been bawling. He had thought I was about to die in front of him. He thought he had lost me.

It was very traumatic. I was in shock for weeks afterwards, and I probably should have talked to someone about it, but you didn't do things like that in those days. It had a profound effect on me because I felt that we could have done things differently and that Dorothy should never have been loaded that way. Thankfully, that scenario would never happen now in Dublin Zoo. Travel crates are brought in advance, and animals are trained to get used to spending time in them. There's a strategy, there's a plan and there's no issue with time. If it takes three months to get them accustomed to the crate, it takes three months.

At the time people asked if we couldn't have used a tranquilliser dart. But it's only in the movies that you dart an animal, and they fall to the ground and asleep in seconds. These drugs can

take 10 to 15 minutes to act because most of them are intramus-
cular. If you're lucky enough to hit a vein, a blood capillary or
an artery, the animal can go down much quicker. But generally,
you go into muscle; then in a heightened state of stress or panic,
like in an escape, animals can absorb these drugs very quickly,
so they are less effective.

Normally, if we had an animal that needed to be sedated, it
would be distracted, then I would fire a tranquilliser into its leg
or bum, close the door and leave it for 10 minutes. The animal
would sit down, with no stress, and fall asleep. In an escape situation, though, adrenalin is high, and decisions have to be made
in a split second.

This was also at a time when the zoo was in negotiations
with the government about funding and support. Had the rhino
got past me, and had there been casualties and loss of human life,
they could have pulled the plug on the zoo. That didn't make me
feel better, but when I thought of the children that I could see a
hundred yards away from me that afternoon, I knew the life of
the rhino was the price for their lives.

Another terrible event was the loss of a mother and baby
chimp in 2000. They had just come to us from Belfast to help
form a troop on the new Chimp Island in the African Plains.
Introducing unrelated chimps is always a bit more complex
than a family group that grows, learns and lives together.
Something disturbed or upset the mother, who ran towards
an area that had electric fencing to protect planting and tried
to crawl under it with her baby on her back. The baby got a
shock on the fence and ran towards the water's edge, followed
by the mother. There are always risks with water barriers and
how they are designed. The Belfast chimps weren't used to

water and drowned. It was horrible. We went out on a boat to retrieve them, and they were still clinging together. It was just awful as another primate to be looking at this dead mother and baby holding on to each other in their final moments. The deaths prompted reviews about using water barriers and moats, and design changes were made.

Harry, Dublin's silverback gorilla, was a shy and subdued animal, but he became one of the zoo's favourites. When he arrived from Frankfurt Zoo in 1995, he was a very insecure nine-year-old who would sit with his hands on his head. He had been separated from his group for quite a while because there were behavioural issues within the group. My father, Liam Reid and I spent hours upon hours sitting with him and trying to bring him out of himself. We finally got him to the point where he was introduced to Lena, and he thrived, developed into a silverback and became reproductively active. However, he was always a very quiet animal. Just like humans, in the animal kingdom, you get extroverts and introverts, a range of different personalities. Harry was happy just to sit around, and he would never display or bang his chest like a typical male gorilla. However, he and Lena were inseparable, and he was great with his youngsters, excelling when they were moved to the African Plains.

Gorillas can live into their 50s, but Harry was 29 when he became sick in 2016. We made him comfortable, and keepers were constantly checking on him and giving him medicine. We hoped and expected him to improve and get better, but he died. The post-mortem showed that he had a brain aneurysm, something we could not have known or predicted. Then, in 2022, there were terrible claims from someone who had worked in the zoo (and repeated in the Senate) that the zoo had been negligent and that Harry had not received appropriate care. They were claims I totally refuted at the time. I would never have stood by and been part of any neglect. I am not sure what else could have been done. To detect an aneurysm would have required bringing him for a head scan, which would have been impossible, because of his condition and body weight. One of the allegations was that Harry had received no post-mortem, but this had even featured on *The Zoo* television programme.

Post-mortems are always carried out in the zoo when the cause of death isn't obvious. A team will come in or we would go to UCD, depending on the animal's size. It is important to know how an animal dies. We have to find answers, and we need information to share with other zoos. This will go in the Zoological Information Management System reports (ZIMs), an international record-keeping system that all zoos can tap into.

An investigation by the National Parks and Wildlife Service (NPWS) into Harry's death and a series of other claims found the allegations to be untrue. The report even stated that Dublin Zoo 'is an organisation with an outstanding track record in animal welfare management'. Only one of many claims was upheld. This maintained that the red panda habitat was not built to the best practice guidelines of the EAZA. However, the report also noted that Dublin Zoo's habitat is consistent with other zoos in Europe and North America that house red pandas.

Harry may have had a genetic weakness. His son, Kesho, went to London Zoo, but he wasn't reproducing there. They were wondering why there was a lack of interest in the females, and a chromosome test revealed that Kesho had something called Klinefelter syndrome. Humans also can have this condition, where boys are born with an extra X chromosome, and it can be linked to infertility.

Harry was an amazing animal. You remember many animals in your life, but he was one that we all took a shine to. His death brought out a lot of emotions in people, but to say that he was deprived of veterinary care was outrageous. I have witnessed birth and death too many times to count in my career. Nature can be savage, and historically, animals were kept in poor conditions. However, the management teams I worked with as a team leader and operations manager were, to a person, compassionate, caring and committed to providing the very best welfare standards possible towards every single animal in their care. At the start of every day, then-director Leo Oosterweghel asked me if I needed any resources for the welfare or wellness of the animals for whom I was responsible. Every day!

One thing we did learn over the years is the importance of allowing animals to grieve. When Harry died, we let the rest of the troop come in and be with him. It's important for their acceptance of death. A younger male kept pushing at the body, as if to check Harry was gone and to make the point, 'It's my turn to be the leader now, he's gone.' And this was the natural behaviour. Many years ago, when Kongola, an alpha male chimpanzee, died, we let the chimps in to see him. A young male, Andy, came up to his body and hit it to make sure he was dead. Immediately Andy's role had changed: he was now the top dog. By allowing him to witness the death of the leader, the natural progression of the group continued. We allowed them to mourn Kongola's death and to understand that he was gone.

Before, when an animal died, it was taken away. We had no understanding of the consequences that action had socially, emotionally and for group dynamics, as they were left wondering where the animal was. Now, we know that the best philosophy is to allow them to have time with the body. Intelligent animals need to mourn. Judy the chimp carried her dead baby around for three days, trying to get a response from it. She only left the body down when it started to decompose. Separation was on her terms. In the wild, elephants are recorded coming back to the spot where a herd member died, touching the bones and spending a couple of hours standing there. They are grieving and remembering. There's now growing recognition of the need to protect the rights of apes because they too experience pain and suffering like humans. It is not just a human privilege to feel peace, to cry, to celebrate and to mourn.

Kavi and Ashoka

Bull calves Kavi and Ashoka and their half-sister Samiya were the first calves to be bred and born in Dublin. They arrived in the summer of 2014 and gravitated towards each other constantly for the first few years until playtime got a little rough for Samiya, and she began to opt out.

When bull elephants are older, they may need to fight for dominance and rights. Kavi and Ashoka may not have known what was in their future, but they certainly loved to play fight. As they got bigger the games got rougher. They pushed and shoved and tussled. We would spot tusk marks and scratches on their rumps. They were so boisterous that we would often get calls from the public telling us that the elephants were fighting. But they were doing just what they have evolved to do – and there's always going to be plenty of noise and disagreements in a herd, just like in a human family.

Gradually the pair started to hang out more on their own, moving away from the herd. Keeper Brendan Walsh was doing a sleep study and looking at the coalitions in the group: who was sleeping beside whom, and how long they were spending together. He saw that Kavi and Ashoka were drifting off together regularly and having boy time. They'd move into the house when the others went outside, or wander back outside when they came in. Just like they would in the wild, these young bulls were separating from the herd, getting ready to move away. Kavi had been a mummy's boy, while Ashoka was more independent, but their mothers Yasmin and Anak had already shifted their priorities and attention to younger calves. The bonds were breaking.

As part of the international breeding programme, it was decided that our Dublin boys should move to Australia to begin a new bloodline of elephants on that continent. We knew they had the education to be good bulls, information they could tap into when needed. Kavi is like his father, calm and steady, while Ashoka is more playful. However, they were both sociable. They had both seen Upali mating and they had both witnessed births. They had learned so much already from their herd.

At the start of 2020, they left for Australia. Their luggage included boxes of apples, bananas and melons, and their favourite playballs – familiar objects always help in new environments. Because of COVID-19, they spent longer than expected in quarantine in the UK. It was December when Kavi and Ashoka eventually arrived at Sydney Zoo, accompanied by Dublin keeper Ray Mentzel. They were greeted with excitement by a new friend and matriarch: Saigon, an old circus elephant. She had not even seen another elephant for a decade, and the company of our exuberant Kavi and Ashoka enriched the final years of her life.

SEVENTEEN

PLEASE DON'T FEED
THE ANIMALS

As a trainee keeper, one of my jobs was to sweep up bin loads of nutshells from the chimps' pit. People bought bags of these outside the zoo, and the chimps would be begging all day. Even then I thought, 'This isn't right.'

There were a lot of good people who came to visit. But very few had any respect for the animals. Many regarded them as stupid and just came to mock and laugh at them. There was a culture of coming with bags of apples or other food, or even rubbish, to throw at the animals. Visitors would throw food at the elephants, or bars of chocolate to the chimps and gorillas. Some people felt feeding the animals was part of their day out at the zoo, others just came for mischief. It was awful. Even when there were signs all over the zoo telling people not to feed the animals, they still did! I'd go into the chimp pit in the morning with Frank Burke, and we'd be filling black sacks with peanut shells, cans of Coke and litter.

People would even throw lit cigarettes at the chimps and orangutans because the animals would mimic them, and they would be puffing on the cigarettes and blow out smoke. People thought this was hilarious. One time Leonie set her hair alight with a cigarette. The zoo had to place somebody on permanent duty to stop people feeding the animals and throwing in cigarettes.

It was a desperate problem. I remember pulling a couple of penguins that had suddenly died. When we cut the poor things open, they were full of little plastic straws that people had thrown into their pool. The zoo banned straws soon after that.

Linda the hippo, one of our most loved animals, had come to Dublin as a baby in 1975 and had seven babies herself. In 2002, she died after swallowing a tennis ball that had been thrown into her enclosure. She must have thought it was an apple and ate it. Nobody reported the ball being thrown in or seeing her eat it; it could have been a kid who threw it in without thinking. The ball blocked her intestine, and as the gas built up in her body, Linda just kept swelling up. She must have been in excruciating pain before she died. I will never forget her suffering. Her keepers Joe Byrne and Ken Mackey were deeply traumatised by her death, as were the rest of us. It was so distressing. Linda's post-mortem revealed rings and rings of burst blood vessels all the way down her intestine because she had been trying to force the tennis ball to pass through her system. It was horrific.

Nowadays it is very rare to see anybody try and feed an animal, or even throw rubbish on the ground. People are a lot more educated about what is happening in the natural world and have more knowledge of, and respect for, the animals. People feel that it's their zoo now, and they're proud of it. Even children have more respect and an awareness of the global issues affecting our planet. For decades, school tours would be let loose in the zoo – 'Be back at the gate at one o'clock' – and it was mayhem. Now they have to be supervised, there is a structure and procedure they have to follow and a culture of discipline and respect.

However, what we were feeding the animals wasn't always appropriate either. There were no scientific diets then. I remember two sisters who lived around the Howth area – I think they were called Angela and Frances – who were very dedicated to the zoo. In their spare time, they would collect waste food from Superquinn shops. They would come in a couple of times a week with food that had been discarded: cut-up or damaged fruit used in displays, and stale bread and sandwiches that couldn't be sold. The keepers would go down to the yard and open up their car to collect the food and it would be shared among the animals. It was seen as a treat for the elephants or a supplementary food for the primates. It wouldn't be acceptable now – both from a nutritional point of view and because of the number of people who had handled that food, and the risk of cross-contamination. But back then the zoo was so poor we had to take those donations.

I would take a couple of bags of bread for the elephants or the chopped fruit, and I would just empty everything out. Elephants don't eat meat and one time I gave them a bread roll with ham in it. Kirsty took off the top of the roll, peeled away all the bits of meat and then ate the bread. She figured it out herself.

We would also get donations of chickens or meat that was out of date, and leftovers or old fish from Howth. I remember as a teenager going out to Howth and emptying fish off the trawlers for the sea lions. We would load them into boxes and bring them back because the zoo was constantly running at a loss. The zoo was in survival mode, and part of my role as a junior keeper was to try and help it survive. I used to go down to Portlaoise once a week, to go shooting with a relation of mine, John Styles. We would shoot maybe 30 or 40 rabbits, and I'd bring them back up in the back of the car for the lions and big cats, just to give them a bit of variety.

In 1997 we lost an entire colony of penguins except for one male, Percy. Initially, it was thought the 11 birds had died from toxic paint poisoning (their enclosure had just been painted), but a second post-mortem revealed that the cause of death was most likely food poisoning, possibly from bacteria created during the freezing process.

Today there are extremely strict guidelines about the fish and meat used for feeding. The zoo uses a lot of horse meat, and that is passed for human consumption. Everything is certified, and we also add carnivore supplements. We never take in random carcasses, and even though there are a lot of deer in the adjoining park, their carcasses are never fed to animals, because of the risks of bovine tuberculosis.

When we feed our animals has also changed radically over the decades. In the past, the big cats would come into their house around half past four or five o'clock. They'd be locked in and get a slice of meat on the bone. Now the big cats have random feeding patterns or bulk-feeding patterns. In the wild, a bulk-feeding lion can eat one hundred pounds of meat and then sleep for days to digest it. So we started following that example and mimicking natural feeding patterns: we might bulk feed our male lion 25 or

30kg of meat with a lot of bone, and he won't eat again for two or three days. And the animals have just excelled. We feed them meat with bone, we tie it up and we have feed poles, so they have to exercise and use their claws to rip it and tear it.

Feeding is used to stimulate and enrich the animals' lives. In 2016, when Dublin's orangutans were moved to a new habitat with trees and high ropes, and they got to do what orangs like to do (which is climb all day), you could see how their physical and mental wellness changed. The habitat includes artificial trees, and in the middle of those is a lift for food. The keepers pack the food in the morning, and the lift is timed to go up and disperse food out into the branches through puzzle feeders. So the orangs are exercising their brains and keeping busy as they forage.

In the old days, we might give the lions a bit of frozen meat in the summertime – sort of a meaty ice pop – and they would sit there for two or three hours with it. Now we might take a piece of bedding from the Zebra House and make a trail through the grass in the lion or tiger area. The big cats follow the scent, like a hunt, until they find a piece of meat hidden under a log. The chimps and the primates like to pick, so we scatter sunflower seeds through the bark chippings in their habitats. They will search happily for hours, compared to the old days, when they had their hands out begging from visitors all day.

When the elephants were fed hay on the ground, it would all be gone in an hour. Now we put it up in a net, which is winched up high, and it prolongs the feeding pattern because they enjoy weaving the hay out of the net. In the summertime, when the browse on the trees is at its most nutritious and plentiful, they have saturated browse days, where we might plant 200 trees in the ground for them. This reflects feeding fluctuations in the wild, where it's not breakfast at eight o'clock and dinner at six.

The horticultural team does an incredible job. Not only do they make the zoo look good as visitors wander through its paths – but they also ensure plantings for each habitat are appropriate and non-toxic. Stephen Butler, a wonderful horticulturist for almost four decades at the zoo, wrote a book called *Gardening for Gorillas*. It's all about figuring out what animals eat and don't eat, and choosing the right selection of plants. Some of the plants are harvested – by animals and staff – for feed. Stephen's legacy will live on for years in tree and plant form.

There is also a plantation of sustainable trees growing down in Lullymore, in the Bog of Allen. These are used for giraffes, okapi, bongos and elephants, though even the large apes love eating bark. Paul Gahan at Lullymore Ecology has managed the zoo's browse programme for over 10 years. He does incredible work selecting tree pieces. He'd often ring me up and say he had a nice piece that might suit a reptile or a big cat, or a root ball for the elephants. It was such a great asset having his eyes to pick out what we would describe as 'furniture pieces' or 'enrichment pieces'. He comes up regularly to decorate and rejuvenate areas like the South American habitats.

Fresh browse is a wonderful resource, particularly for elephants, who should always have access to it. It's critical for how they live, how they feed and how they interact. When we were developing the feeding programme in the zoo, we used to put a lot of browse in the sand and the indoor floor, then the elephants would come in and knock it over quickly. So we started soaking it. The elephants didn't want to knock it over when it was wet, because the sand would attach to it, and they would spend hours breaking the branches rather than letting them hit the ground.

Elephants have their preferences in the types of browse they eat. They don't like poplar, and they don't like sycamore for some reason. They particularly like birch and willow, and they love bamboo. When we started putting sea salt, vanilla or different spices on the browse, they would smell and choose one over the other. Different elephants like different things. They all like sea salt. But some like vanilla, and others prefer herbs and spices. Some love garlic, while others won't touch it. So even with the same species of browse, they have the choice to eat their favourite first.

In the wild, elephants see hundreds of different variations of plants. An adult can eat between 2 and 5 per cent of their body weight daily, which is a lot, considering their size. What they eat – hay, bark, leaf material – is high in fibre and low in nutrition, but it is turned into protein by an enzyme produced in their body. Their teeth are designed to grind down this fibrous diet, they can eat branches as thick as your arm. Sometimes it's said that an elephant has four teeth, but this isn't really true. They have over 100 teeth, fused together in blocks that grow throughout their lives in a process known as a molar eruption sequence. They usually get seven sets of teeth in their lifetime (we'd often find a big lump of tooth); in the wild, after the last set falls out, they will die because they are unable to eat.

The digestibility is about 40 per cent, so elephant dung is practically hay coming back out again. That makes great fertiliser, so it is used for growing more plants at the zoo and in Lullymore; Dublin City Council uses it on planted areas around the city.

We now understand the complexity of animal diets. Years ago, in the old Monkey House, there were monkeys from South

America, lemurs, African monkeys and Asian monkeys. In the past, they were only fed fruit. (I remember going in as a kid to chop apples and bananas, to throw to them.) Now we have pellets based on what the animals would eat in their native region. The okapi is another great example. This forest giraffe, which was only discovered a little over 100 years ago, has very particular requirements. We even bring in roses for them at certain times of the year, and we have to supply different variations of browse for every feed because they'll decide they like one, but not another.

The primates love eating nutritionally rich insects, and they enjoy catching them. In the past they only got the cockroaches that fell into their cage; now they get locusts, crickets and mealworms. A mealworm is like a vitamin capsule for a monkey because they are fed a specific high-nutrient diet. The worms might be put into overhead pipes with holes so they drop to the ground randomly, and the chimps or monkeys start scurrying to find them.

There have also been changes in what we give to animals to drink. The apes were occasionally given Guinness, when they wouldn't take medicine in anything else, or if they needed a tonic. We had a very large male chimp called Kongola in the 1980s. He had the flu and wouldn't take his prescribed antibiotics, but he downed them when they were finally given to him in a bottle of stout. A short time later, he and Betty had a baby, Vicky, and one day another keeper and I came in to find Kongola with the baby in his arms. He had taken it away from the mother. The baby was becoming very distressed, and we didn't know how we would get her away from him. If we had called the vet down, that would have excited the whole group, and Kongola would more than likely have bashed the baby off the wall. So we decided to go up to the bar and get him some bottles of Guinness, to which we added a little oral sedation. We gave them to him one after the other until he relaxed, and then we went in and took the baby out. Vicky spent a little time being checked over in the Rotunda Maternity Hospital before the head curator at the time, Ron Willis, took her home to hand-rear.

We no longer take animals away from their families for hand-rearing, but when we began assisted feeding elephants, we would use human infant formula. It was very easy to get, but it does not match up to the nutritional requirements of an elephant calf, particularly in terms of fat content. Now we use a milk that was designed in Germany, based on samples from lactating elephants. It's a mixture of powder and fat and includes a lot of coconut oil and rapeseed oil.

There are now special diet services around the world making specific animal foods. It's very complex and expensive, but we know that diet has a direct impact on animal health and animal management. The zoo now has Andy Beer, formerly of Sparsholt College and an expert in zoo nutrition, to advise, and the veterinary team will also be involved in diet management. The zoo also works with nutritionists and uses international software that provides diet data to zoos. This allows nutritional management planning based on very precise observations of an animal's movement, condition, age and other factors.

Nutrition and feeding strategies are important because animals in human care don't have the choices of the wild. So it's important that you give them variation and prolonged patterns over a 24-hour period. If they know they are only going to get fed in the evening or the morning, it can affect their behaviour. They're waiting, and it can affect social play.

What I emphasise to any zoo I go to is to get out there in the morning and get all their resources ready: the mud wallows made, the overhead feeders full, the feeding pipes in the ground, the browse planted ... Make sure all the opportunities are there when the animals come out. In a lot of zoos, the elephants might be having intimate social interactions, when all of a sudden, a keeper brings them in, so more food can be put out. We don't know what's going on in those moments, but they're communicating with each other, and we shouldn't interrupt those moments by not being prepared.

Zulu, Jabali and Kamili

In November 2022, when I first met Zulu at Al Ain Zoo in Abu Dhabi, she charged full steam at the fence line trying to attack me and trying to grab her keepers.

I understood why. She was just reacting to the situation she had found herself in. She and her seven companions had been taken from their home in Namibia's wild and sold to this safari park. She was in an alien environment.

As the oldest elephant, she had lived in the wild the longest and had the most to lose. All she knew, all her wisdom, was of life as a wild African elephant.

I was told she was 40, but I'm not sure how that age was deduced. She was about 2.6 metres tall, in great physical health and probably weighed about three to four tonnes. Her right tusk was shorter than her left, but this may just have been natural wear and tear because she had a preference for using her right one. Just like humans have a preference for using one hand, elephants have a tusk preference, right or left. You often see where the tusks are slightly uneven or worn down on one side.

The group around her looked like a challenging mix, and it's unclear if they were related. There was a lot of unity within the herd, but it was probably through adversity that they gravitated to one another. However, Zulu was their obvious leader. She had allowed two of the calves to suckle, which showed her maternal instincts.

All the elephants I have worked with throughout my career have been accustomed to people, and have responded to people. Zulu was different. All her experiences and interactions with humans to date were likely to have been negative. She was cautious and full of attitude.

When we started training and she realised I was offering treats, her body posture relaxed a little. She wasn't trusting, but she was interested. She remained wary for the first few days, which was completely understandable – as the matriarch, she felt responsible for keeping the herd safe.

However, she quickly showed that she was a clever, steady elephant. Gradually, she was willing to move forward towards the keepers and showed plenty of potential, though gaining her full trust was likely to take some time. She also wanted to control, and she lashed out with her trunk when she wasn't being fed quickly enough.

She seemed intolerant of the two young bulls, but this could be because she was related to them and needed to drive them away from their familial group, as she would in the wild; alternatively, she might have been unrelated to them and seen them as extraneous to her family group. Genetic screening would be needed to show who was related.

One of the young bulls seemed particularly intimidated by Zulu and was reluctant to enter the fray when looking for food rewards. He had been called Jabali, which means 'strong as a rock'. I was told he was 12, but he looked quite small. He got pushed around a good bit by the other elephants, and he seemed so conscious of the other elephants around him that, initially, he was unable to relax enough to focus on the keeper and target during training. I told the team he would need to spend time away from the herd with Epesi, the other young bull, to build up his confidence.

We weren't quite sure about Jabali's relationship with the other herd members. Masego or Zulu may have been his mother, but we won't be able to confirm that until we can genetically screen the herd. However, seeing some pushing out or aggression towards him was very normal behaviour. He was at the age where, if he was in the wild, he would start to fend for himself or join a young bachelor group and live on the periphery of a herd. The best way of managing the situation in human care is with what we call fission–fusion. We gave him time alone, or with other young bulls, to develop his confidence and then some time with the herd. So he might spend 48 hours with the herd and a week away from them. With time he will get more dominant, and then they will show greeting behaviour to him and accept him more. Fission–fusion means he will drift in and out for positive interactions and learn to exert his dominance.

However, when I began training with the herd, Jabali was one of the first to figure out what was required of him. He was very smart. He loved his daily shower, a wash down with a hose.

He would run towards me, ears flapping, looking to interact and learn. He also seemed drawn to Kamili, a female about 15 years of age. When we started training as a group, Kamili was one of the most cautious elephants. She preferred to hold back and watch what was going on before joining in. She would come forward and take rewards, but from the start, I could see we had some work to do to build her confidence.

Kamili's distinguishing characteristics include a big tear in the middle of her left ear. The most obvious difference between African and Asian elephants is the size of their ears (African elephants' are bigger). There is a whole series of veins behind these remarkable appendages, and when they flap it cools the blood down as it passes back into the body. African elephants are also more wrinkled, and that helps to retain moisture and mud to help them control and modify body temperature.

I realised that we would get more focus from Jabali and Kamili during training if they were separated from the group. And as soon as we put them together in a stall at the furthest end of the house they became more at ease. They appeared relaxed in each other's company, spent time outside feeding together and worked well together, too.

At the beginning of one session, Kamili appeared agitated and was reluctant to come into the house. She kept looking up at the ceiling and secreting from the temporal glands on the side of her head, a sign that she was stressed. We finally realised that the air-conditioning unit had been left on, and the noise and vibrations were distressing her so much that she didn't want to enter the house.

I decided to explore gradual and controlled ways of desensitising Kamili to the air-conditioning sounds, a sort of exposure therapy, including giving high-value rewards like large browse pieces when the air con is on. In the meantime, signs were put up reminding everyone to ensure that the air-conditioning unit was turned off before the elephants were brought in to train.

I wondered if Kamili associated the air con with being darted from the air by helicopters in Namibia? But regardless of the reason, it reminded me that these are animals that have experienced significant trauma in their lives. Trauma for elephants, like people, is very complex and can reveal itself in unpredictable ways. We must always look for reasons and solutions when an elephant shows signs of distress.

After six months of working with the herd in Al Ain Zoo, the difference is extraordinary. The elephants are all training really well and the zoo is putting massive resources into making sure they have a life of autonomy, choice, opportunity, control and decision-making. The safari at the Al Ain Zoo will be one of the world's best and biggest. I know it's not the wild, but at least with me there as a consultant, defining and planning their lives, I can personally make sure these elephants have the best care possible.

EIGHTEEN

ANIMAL HOSPITAL

Nature never ceases to amaze me. We train many animals for veterinary access, but they still don't like needles. In certain animals, we might even have to use sedation to get a blood sample, and that reduces the flow of the blood. One stress-free solution is to use 'living syringes', such as Mexican kissing bugs, to get samples. If we introduce these insects – also known as triatomine bugs – onto the back of an animal, they send out a probe that emits a slight anaesthetic before they insert it through the skin or hide. They can detect where the blood capillaries are, and they draw out the blood. Their abdomen will be pancake-flat at first and then swell up. They are incredible.

Imagine trying to figure out how to get a blood sample from

a porcupine, with all its quills! André Stadler from Innsbruck Alpine Zoo in Austria, who introduced the bugs to us, was able to put a little harness, on a thread, on one. He placed it on the quill of the porcupine, and the kissing bug would walk all the way down, latch itself onto the skin of the porcupine and take a blood sample. After a few minutes, André would pull it out slowly, and the bug would have a belly full of blood. It was an ingenious idea.

We use a syringe to extract the blood back out of their abdomens, without harming the insect. Each bug extracts about half a millilitre of blood, so we usually need about four bugs for a sample. The bugs are specially bred in Wuppertal to be pathogen-free, and they are kept to be used again on the same animal. We first trialled them back in 2007 to check hormone levels in Yasmin the elephant, who was pregnant. At that stage, the herd hadn't been trained for veterinary access. Since then, we've used them as much as possible to reduce stress levels in animals. They've already helped to check for TB in sea lions, hormone cycles in rhinos and health screening in tapirs, who would normally have to be sedated.

When I began, the vets only came in a couple of times a week or whenever there was a problem. We would bring the animal to a room where lots of things were stored and cleared off the table if we needed to do anything with it. It is no surprise the 1990 investigation into the zoo criticised the lack of adequate facilities for housing or nursing sick animals.

Our vets remained part-time for much of my career but committed experts like John Bainbridge were always on call when we needed them, and available at the end of the phone for any medical question. Today there is a full-time veterinary team on site that works with vets all over the world, exchanging information; we also have a full hospital, where large-scale procedures can take place. Animals can be sedated and brought down to the hospital, and specialist vets and veterinary colleges come to us now if we need support. For years, when an animal needed medical care, I'd often have to catch it, put it in a box and take it across the city to UCD Veterinary Hospital – sometimes under police escort – and then back again. It just added to its stress.

Since I began as a keeper, there have been many advances in medical care for animals. We have a long history of working with the veterinary school and hospital at UCD. Equine vets are very good at treating zebras and other ungulates, and the system of a cow is very similar to that of some of our herbivores. Many of their specialists have developed their skills working hand in hand with the zoo – for example, Dr Vilhelmiina Huuskonen, a lecturer in veterinary anaesthesia, has now honed her skills on at least 12 giraffes, a very difficult animal to sedate safely.

Before sedation and anaesthesia were perfected, there were times when I would be in the same space as a big cat for medical reasons, and they would suddenly stand up. I would use a yard brush to push them aside and jump out of the way. Sedation and anaesthesia, when necessary, are better now, and a lot safer for both animals and humans.

It can often be hard to determine when an animal is getting sick, or what might be causing discomfort. Sometimes the behaviour of other animals can tell us something is wrong: a gorilla staring into or smelling another's mouth could mean the presence of an infection.

For most species, faecal samples are checked once a month for pathogens or worms, which are common causes of illness. If an animal is sick, the protocol is to get a temperature, faecal sample, blood sample and weight before doing more invasive tests, because it could just be worms or a bacterial infection. Where possible, we train animals to accept regular veterinary checks so we can monitor their health.

Wanita and her brother Emas were two Sumatran tigers born in May 2009. Theirs is a critically endangered species – there are thought to be only hundreds left in the wild. The area where they come from has seen catastrophic deforestation for palm-oil production, which is a common ingredient in chocolate, biscuits, processed foods, soaps, cosmetics and cleaning products. That's why we all need to make conscientious decisions about what we buy. Part of caring for our planet and animals is checking out that you are not supporting the palm-oil industry. In Sumatra and Borneo, hundreds and hundreds of acres of trees, where orangutans could climb, are being deforested every week. In the course of one hour, an area the size of 10 football pitches will have been excavated for palm oil. This needs to stop because there are plenty of alternatives. In 2019, the city of Chester in the UK said 'Enough is enough' and became the first Sustainable Palm Oil City in the world. More than 50 organisations committed to sourcing palm oil from entirely sustainable sources. We need to follow their example.

One of the smallest subspecies of tigers, Sumatrans are beautiful. They do a lot of their killing in water, so their feet are almost fully webbed, which allows them to swim and pull prey back out. They are such a perfect evolutionary specimen for what they are designed to do, which is to hold, puncture and kill. Even though I'm associated with elephants, my history began with big cats. I've always been in awe of their power and strength.

The cubs were the direct result of an international animal breeding programme, with Kepala coming to us from Chester

Zoo. He was a magnificent tiger and genetically very important because he had never been represented in the European population. It normally takes six to eight months to introduce a male to a female, but Kepala arrived in early February, we put him with Sigra on 13 February, and they mated on Valentine's Day. It was love at first bite! The result was Wanita and Emas. Traditionally, male Sumatran tigers prefer a solitary life, but Kepala took an active role in his cubs' upbringing and even slept alongside them at night.

We noticed that Wanita seemed a bit more lethargic than her brother, who was a boisterous little chap. It was a cause for concern, so we let her mother outside for a walk one day and gave Wanita a full medical check-up. Vet John Bainbridge became concerned about her heart rate, saying the beating didn't seem quite right. We took Wanita to see UCD's cat specialist, Barbara Gallagher, because the physiology of a tiger cub isn't that different from a domestic cat, and it turned out that she had two holes in her heart and other cardiac issues.

We were told that she probably would not be able to become a mother because of the strain the additional weight would put on her heart. However, it was suggested that we start doing some physiotherapy with her because, in addition to Wanita's chest being flat, her front legs weren't as strong as they should have been because she wasn't moving around enough.

Based on Barbara's recommendations, we put a physical fitness regime in place to build up her chest and her front area. I would go in with her on a daily basis with other keepers, and we would lift her up off her back legs, so she had to put weight on her front legs. We would just let her walk forward and back, forward and back, on her front legs for five or ten minutes. Gradually we could see her chest shape improve – it was no longer flat – and she began to catch up physically with her brother.

Further medical checks showed that a heart murmur had gone and one hole in her heart had closed up. Her remaining cardiac issues were small and not expected to impact hugely on her quality of life. However, as she got bigger and stronger – and more physical – Wanita still needed regular checks. We didn't want to sedate her, so we had to come up with another way of accessing her heart. I started a training programme: standing on the other side of the mesh fence, I got her to follow a target stick and rewarded her with little pieces of meat. And she took to it. She would stretch upright on her two back legs – she was now so big we were eye to eye – and I trained her to push her chest against the mesh. I would train with her almost every day, but once a week the vet would come in to check her heart. We designated an area out the back with a larger hole in the mesh, and she would follow the target stick, push her chest forward and hold steady in position, so the vet could listen to her heart with a stethoscope.

Wanita was a big favourite because she starred in *The Zoo* TV programme. However, it's not her training that most people remember. When I brought her out to the UCD vets for a check-up, I was in my green uniform, but she defecated all over me when she was being given an injection (as animals tend to do), so they gave me a set of blue scrubs to wear. As I walked out of the surgery, the cameraman caught me saying, 'Went in in green, came out in blue, and all because of tiger poo.' After the episode was broadcast, people would be chanting that at me in the supermarket.

Wanita herself became so well and fit that she went on a breeding loan to France. We didn't want to risk sedation, so we trained her to travel in a crate, spending weeks preparing her for her journey, both physically and psychologically.

Animals regularly need dental care, too. Inda was another tiger. He had been kept by a private owner, a merchant seaman in Germany, I think. He was found in the back of a house with all his teeth worn down from biting on bars. His canines and carnassial teeth were all damaged, and we could see that he was probably in pain. A carnivore's teeth are essential to their survival, so we flew in Peter Kertesz, a renowned London dentist from Harley Street. Although he is a human dentist, Peter now specialises in supporting animals, and we regularly use him for dental and tusk care. Inda got a root canal treatment, fillings and a new lease of life!

Keepers also get excited about first poos or bits falling out of animals that we know have been causing pain and discomfort. Other people would cringe at the strange situations that make keepers happy, but we are delighted to give any kind of relief to a living creature.

Asha, the first elephant to be born in Dublin Zoo, retained her placenta when she gave birth in 2016. She was given oxytocin (the drug sometimes used during a birth process or to help release the afterbirth), but that didn't work. You could see bits of the afterbirth hanging from her, and the young bull calves would come over and break pieces away. We were really worried that the placenta would start to decompose and that Asha would get an infection in her uterus, so it was essential that we got it out. We brought her into the stalls every day for hot showers, washing around her vulva and below the tail to stimulate blood flow and to try to encourage her to push out the placenta.

As a keeper, I'll adapt anything at my disposal to make an

animal's life easier. For Asha, I found a clean litter picker. When she was warm, and her body was a bit more relaxed, I would reach in and slowly start to pull the placenta out using the litter picker. It didn't cause her pain – indeed, she was quite happy for me to do this because it was a separate entity from her physical structure. I kept manipulating and pulling, and eventually, it worked: one day, out came the placenta. We were all relieved.

Modern keepers always consider the rest of the family or group when an animal needs medical attention. When an animal has to be removed for medical care or surgery, we try to return them as quickly as possible. Keeping a chimpanzee, in particular, out of its group for even a short period of time can change the social hierarchy very quickly and could even result in them being ostracised.

I've seen chimps, wolves and hunting dogs with significant bites and injuries, but that's all part and parcel of dominance hierarchies. There will often be aggression because that is what happens in the wild. However, in the wild, animals can go off and heal at the periphery of the group and come back in. In a zoo situation, if you take them out for too long – whether hours or days – it could have long-term consequences: they might never be accepted back, or they might be attacked and even killed.

If the boss, or a high-ranking female, is injured or needs medical care, you can take them out of a group, pride or pack, and there will be a celebration when they return. However, if it's a lower-ranking animal, the chances of him going back in for any reason are minimal. Once he's gone, he is disregarded. Therefore, we have to make a very careful evaluation, taking into consideration the keeper's experience with the animal and the vet's advice. The animal may be in discomfort for a couple of days, but long term he's going to be kept within the group. It would be common to see hunting dogs and wolves giving each other significant bites at a particular time of the year, and then to see them rolling around playing together a few weeks later. It's part and parcel of the turmoil of social adaptations and social structures within animal groups.

That is why training animals to give access to veterinary care is so important. If they get a bite or a scratch, they don't have to be removed. They come over, you give them a treat, and then you can put some topical application of antibiotics on the skin or give them an injection. Training is a vital part of modern animal wellness.

NINETEEN

MY LIFE AS A BOXER

Boxing was, and still is, another of my passions. If I hadn't become a zookeeper, I could have been a contender!

When I started as a boy, nobody had their own gear. There would be a big box of gloves in the centre of the gym, and we would grab a pair. There would always be a run to grab the best set that hadn't been flattened too much, but even the best ones always stank! We must have been boosting our immune systems from the exposure to bacteria every time we put a pair on or put a lace in our mouth to tighten them.

As I got older, it was suggested I go around to the Phoenix Boxing Club, because there were more guys my age. It was a small club around the corner, on Parkgate Street, and we'd almost be in the ring as soon as we got inside the door. When we sparred, we couldn't go back too far, or our heads would hit the wall, which was splattered with blood from where fellas had hit their heads or from punches. The smell of the chipper next door would be mixed with fresh body odour and stale sweat. It was an antiquated gym, but it produced some great champions.

Tony Davitt, the legendary Dublin boxing coach who's still going strong in his late 70s, was the main man, and Peter Glennon and Sean Wright were there as well. Tony used to bring down his brothers PJ and Tommy, who were pros with Eastwood's Gym up in the north of Ireland, to spar with us. We were only kids, about 15, and sparring pros. If you dropped your hand down, you might get a left hook that meant you wouldn't eat for a week after. It was a great learning curve because when we got out to fight fellas our own age, it all seemed much easier! But even sparring with other teenagers was tough. Sometimes the sparring sessions were harder than the actual fights, even with fellas who were lighter than me. The sparring and the competition were fierce, but the camaraderie was great.

Stevie Collins, who went on to become world champion, used to spar with me a lot. He was Irish junior champion and I was the Irish youth champion. We still meet up in his brother Paschal's gym in Corduff, where I bring my son Zac.

Tony Davitt's methods to make you do better were unique. I was fighting in the finals at the Dublin Leagues and thought I'd had a great first round. When I came back to the corner, Tony said to me, 'Do you know him? You're only effing short of kissing and hugging him. Why don't you go out there and give him a bigger kiss?' That was Tony's way of telling me to get out there and do better. He was great, an amazing man with an incredible dedication to boxing. He travelled up and down from Stamullen in County Meath to train us three times a week.

All our trainers were really dedicated: they gave up their time and took us away on trips. There was a Dublin versus London event every year when we'd go over to the Repton Boxing Club and fight in the Barbican Centre. It was the 1980s, and there were a lot of Irishmen doing well there in the building industry. It was a big dicky-bow affair. We'd be brought along and paraded out. With my reputation for knockouts, all these businessmen would be asking if I was going to win, and I'd say, 'Yeah, I'm going to stop him in the second round.' And if I did, they gave me 30 or 40 quid, which at the time was like winning the Lotto.

We would stay in an old seaman's club right beside the Royal Oak pub. It's closed now, but above the pub was a boxing gym, and it was where Frank Bruno, Gary Mason and other pros trained. They would let us come in and train beside them, using the bags. Bruno was an incredible athlete and a massive man: his abdominal muscles were like blocks. And we'd go up to Charlie Magri's shop to buy our stuff. (Charlie was a former top professional boxer and a world title contender.) We were like kids in a sweet shop.

However, they were tough days financially. Boxing was only starting to get its act together. We'd be driving around the country to matches, and we'd be asking for a tracksuit, a pair of runners or a voucher instead of a trophy. We had enough trophies to last us a lifetime, but we didn't have the money to buy new gear.

There might be 20 of us in the back of a van going up to Belfast for a boxing match. At the height of The Troubles in the 1980s, we'd go into boarded-up clubs in Loyalist areas, where the door would be barricaded behind us. But religion meant nothing. You got in the ring, boxed together and had lemonade together afterwards. They would escort us out of the estate in their cars. The word would go out that we were boxers up from Dublin and nobody was to interfere with us. Boxing crossed the political divide, and boxers like Barry McGuigan brought everyone together.

I travelled all around Europe with the club and represented Dublin, Leinster and Ireland at internationals. My first international for Ireland was against Wales in Barry. I was so proud of representing my country, of wearing that green jersey, that I wanted to sleep in it. And as the captain of the team, I got to meet Henry Cooper.

The highlight of my boxing career, however, was representing Ireland in a 12-nation tournament in Raufoss, Norway. The team won three golds and four silvers – I think we were one of the first international teams to win gold medals. I was also chosen as the best boxer of the tournament.

Our coach was Johnny McCormack, and I will never forget his approach. The fellow I was boxing in the final was doing very well, and when I asked Johnny for advice, he casually said to me, 'Jaysus. He has two hands, a head and two legs, the same as yourself. It's only up to you to use them better.'

Gradually, I started seeing changes in the sport and the approach to training and looking after the well-being of the boxers. I remember the first time I had to use headgear. In those days, there was a buckle, and it got embedded into my chin when I got hit with a hook.

It was the Cuban trainer Nicolas Cruz Hernández who brought the science to Irish boxing. He was amazing. At that time it was all about getting in there, bating the head off them and seeing who fell first. With Nicolas, we'd use our balance, our rhythm and our strengths. He individualised training programmes for each boxer based on their skill sets.

I was in my early thirties in 2003 when Nicolas was training Billy Walsh to represent Ireland in the Boxing World Cup in India. Me and Dennis Galvin, a really good fighter from Moate Boxing Club in Westmeath, were brought down to Wexford as sparring partners for Billy. Nicolas would have us rolling head over onto our backs for 30 seconds then jump up into a stand. We would be really dizzy, but he'd say, 'Put your hands up, fight, fight,' and he'd come at you. He was showing us how to cope after a hard shot. It was a brilliant way to learn how to compose yourself under pressure.

The training was intensive. He'd have us sit on a big tractor tyre, with a sledgehammer in our right hand and a car axle in our left, and we'd be hitting the tyre to build our body strength. And he had us running the beaches at Curracloe every morning.

It was non-stop. We were tormented. Dennis, who was a bit of a messer, said, 'He's going to kill us. We need to get at least one day off. How are we going to do it?' So we put three or four laxatives in Nicolas's breakfast. He didn't move far from the bathroom that morning, and by 11 o'clock, training was called off.

Billy Walsh, who's now head coach of USA Boxing, is a great guy, and a homegrown talent that was let slip away. I was in Colorado for a workshop about elephants a few years ago and ran into him at a bar. He asked me what I was doing, and I explained about the workshop and then an idea occurred to me. The next day I had Billy up in Cheyenne Mountain Zoo talking about high performance and motivation.

Boxing also gave me great opportunities that a young man from inner-city Dublin didn't normally get. I was introduced to different countries and cultures when I went to Norway, Sweden, Denmark, Germany and Italy. I loved doing a geographical check before going, to see what species of animals I might see. When I went to Raufoss in Norway, I was hoping maybe I might see a wolf or a polar bear because they were in that region.

Even visiting London for the first time was an experience – wondering how the Tube worked, performing at the Barbican, walking down Oxford Street … We were all hormonal 17- and 18-year-olds in London, and we'd go down to Soho to look at the windows. There was nothing like it in Catholic Dublin. I remember the coaches, saying, 'Five minutes is all you're getting,' but I think they wanted to go and look as much as we did.

People often ask why I didn't go professional. At that time, you would have had to go to the UK. But I knew guys over there, and it seemed like they were being treated as second-class citizens, that they were only being used as punching bags. The professional game had no appeal, because of the attitude towards Irish people in the 1980s.

However, boxing has played a huge part in forming the person I am today. It has given me discipline, opportunities and the confidence to help myself and other people. Boxing is a good way to build character. I can see it in my son Zac, in how he carries himself and how he reacts to situations. He's been boxing since he was three or four, and he's absolutely flying. He's down in Corinthians, and he's far better than I was at that age, with a great boxing IQ and a lovely style.

It's great to see the evolution of boxing in Dublin, and Ireland is currently punching well above its weight globally. We're a fighting nation. We have a fighting heart. We have these attributes because we've had to do it for many reasons and many years. There has always been talent in this country, but with better investment, better opportunities, nutrition and psychological support, our boxers are now excelling. You only have to look at our women's team, which is the best in Europe. All the boxing clubs are packed, and it's because of the likes of Katie Taylor and Kellie Harrington inspiring a whole new generation.

Katie's also animal crazy, and she's come up to the zoo for a walk around. She's totally in touch with the reality of the threats to the natural world and the role of conservation. It was great to see her interest. My son Zac was there too. He had his club medal, so she put it around her neck and gave him her Olympic medal to wear around his.

I don't box anymore, but I go away with Zac and his club on training camps, which is great fun. I won my first league fight when I was 11; 43 years later, Zac won his first league fight at the same age, in the same stadium. I was so proud of him. I had tears in my eyes watching history repeat itself.

Tammi

Tammi and her daughters were born in zoos, but it is hoped that they will live out their lives in the African wild.

Tammi came to Howletts Wild Animal Park in England from Israel when she was about a year and a half. She now heads a group of 13 elephants, from two interrelated families, that have left Howletts to be rewilded in Kenya.

Tammi always has a protective eye out for her three daughters Jara, Uzuri and Mirembe. She and Jara, the eldest daughter, are particularly close, and they often greet each other fondly, even after separations of only a few minutes. Jara is likely to become the matriarch of the herd after her mother. Inquisitive and clever, she is often the first one to find hidden food – ideal traits for a future leader.

Uzuri is probably the most placid of the group and very affectionate. She likes to hang out with Manzi, Juluka and calf

Oku. Mirembe, her younger sister, is very active and mischievous, and she loves to explore.

Tammi is the oldest elephant in the group; the youngest, Nguvu, was born in March 2020. I look at him and Oku, and I am so excited to think that they are being given the potential to grow into magnificent bulls of six or seven tonnes, free in Africa. In 20 years' time, when I'm an old man, they could be walking across the Kenyan grasslands, having reproduced. I want this to be something that my grandkids will be telling their grandkids about.

We want all of these elephants – over 25 tonnes between them – to have the most stress-free journey possible, so I have been overseeing their training for their 7,000km trip. Formerly reluctant and reticent calves now run into the crates to train. When they hear the gate being opened (which signals training), they sprint towards us. It's a joy to behold. When I see Tammi and the herd hurtling towards me, I tell myself to get inside their head, to think about how good they are feeling mentally and how enthusiastic they are, and I try to mirror their positivity. They are good mentors and great motivators.

When we did mock loadings, a rehearsal for the big day, we brought over the South African team that will manage the transport, unloading and other logistics at the other end. These are experts who have moved thousands of elephants across the African continent – because of population management or re-locating problem elephants that were coming into villages and eating or damaging crops. Kester Vickery, the wildlife trans-location specialist, told me that he had moved up to 3,000 elephants, usually by darting them from a helicopter, putting them into crates and transporting them to another part of Africa.

He watched the elephants training – they didn't realise this was a big rehearsal – and couldn't believe what he was witnessing. One of his team asked how I got 13 elephants to walk calmly into a crate and close the door behind them. 'With kindness,' I said. 'Kindness, patience and preparation.'

TWENTY

STEPPING DOWN

Kavi and Ashoka, two of Dublin's young bull elephants, are what I would describe as elephant gold. They have seen so much in their young lives, watching and learning from their father Upali and other herd members. When they were selected by the international studbook keeper to go to a new zoo in Sydney, Australia, we were very excited for them. They were going over there to represent Europe, specifically Dublin Zoo.

To move to Australia, the elephants had to be trained for travel and quarantined for three months, so they could be cleared for TB and other health issues. Doing this in Dublin, the whole herd would have had to be quarantined, and the area blocked off to visitors. In addition, some very small elephants would have had to be tested for TB, and this would have been very stressful for them. So we asked Knowsley Safari Park in England if we could quarantine and train the elephants there. Although they no longer had elephants, they still had the facilities to house them.

The half-brothers are very close, so we knew they would be a great support to each other. Their journey began in January 2020. We had trained them to walk into crates, which were then lifted onto the back of a low loader and secured. We had a full police escort from the park down to the dock, where we put them on a ferry.

We got to Knowsley and began training them to leave for Australia in 90 days. At any one time, we had three keepers – myself and Ray Mentzel from Dublin, with relief from fellow team members Donal Lynch, Karen Carrigy, Hannah Wilson and Christina Murphy – and we employed another keeper, Albert Pamies Palazuelo, from Spain. In addition, there was support staff at the zoo. It was a quarantine area, so we had to go through a capsule every day, take off our clothes and put on different clothes when we were working with the elephants.

It was all going really well – until COVID-19 arrived, and new quarantine and travel rules were enforced. Countries locked down, and suddenly we could no longer fly to Australia. I was stuck in England. I was able to use the ferry under strict guidelines, but then the risk became so great that I stayed in England and lived in a hotel. At one point, I was the only person staying in the entire hotel, and the highlight of my day was a trip to Tesco to buy groceries. I managed to get home for a short break, but when I talked to Alan Roocroft, who was working with us at Knowsley, I realised he didn't sound well. I returned straight back to England. Alan had COVID-19 and needed urgent medical help. He was taken to hospital and was in for quite a while recovering.

Then I tested positive, and two more pod members were positive too. Training had to stop. I was in the hotel room for 12 days on my own in total isolation, aside from my phone. Food was left outside my door. The hotel kindly gave me some weights to use for exercise, and I tried to structure my day with Zoom calls and conversations with staff back in Dublin. But it was a difficult time, and I found it quite a challenge. Luckily, Ray, one of Dublin's most experienced elephant keepers, was at Knowsley with Albert, and they had great help from Jen and David Southard, who lived on-site and usually ran photography courses in the park.

l thought everything would open up again after a few weeks; however, months later, our small crew of keepers were still taking turns to mind and train Kavi and Ashoka. Throughout 2020, I was staying at Knowsley for up to 10 weeks at a time. With the zoo closed for long periods, I was needed less in Dublin. The staff there had also been broken up into separate pods to try and ensure everyone didn't get ill at once. Keepers couldn't go from one section to the other, and mess rooms sprung up around the zoo. It was very difficult operationally, and people were under a lot of pressure to keep it going. But they all excelled.

None of the plans that any modern zoo has in place – which include preparing for terrorist attacks and emergency escapes – could have prepared us for the pandemic. With the zoo gates closed, many of the staff (excluding our animal care team and some support staff) were laid off. It was very upsetting. All our colleagues in gardening were let go or put on leave, and all the people in the shops were gone. However, as Christoph Schwitzer remarked, 'You can't furlough an elephant', so the keepers needed to be there, though it became apparent that even some of these numbers needed to be reduced.

We had no visitors, but we still had to maintain the diets, habitat care and veterinary care as normal, which equated to about €800,000 per month. The elephant herd alone cost €5,000 a week to feed; the lions would eat an average of 20 or 30kg of goat, costing about €50–60 per feed. And there were specially formulated diets being brought in from all over the world.

The whole zoo was upended, with people working in different areas. The keepers worked tirelessly. We placed particular focus on our environmental enrichment programmes, presenting food in different ways to test the animals' physicality and mental capabilities. However, they could sense the difference. Primates, particularly the larger-brain species, are people watchers and enjoy seeing people walking past. I also missed that stimulation, the buzz of children and the excitement of voices, or the elephant talk at 12.30 p.m. I used to love being there for that: we'd call the elephants over, and there would be gasps of excitement from the children as they came into view. I missed that energy from the public.

We opened for part of the summer, but there were considerable restrictions on numbers. We were allowed to admit only a couple of thousand people. In 2019, we had 1,270,000 people visit the zoo. By the end of 2020, the figure was less than 500,000, and the loss of revenue was an estimated €10 million. We had learned to keep some emergency reserves, following our experience in 2001, when the zoo was closed for over nine weeks to protect any animals that might be susceptible to foot and mouth disease. However, it wasn't long before those reserves were gone.

A financial crisis was inevitable, and the onset of winter and additional heating costs brought the situation to a head. We had been talking about restructuring and redundancies. Now we were discussing what would happen if we couldn't open our doors again. We would never euthanise animals. In other circumstances, we would have been looking to other zoos to take our animals, but a global pandemic meant that many zoos across the UK, Europe and the US were also closed, and also in financial difficulty.

In November, I fronted a public appeal to save Dublin Zoo, asking for even the smallest donation to help us look after the animals in our care and to keep us open. People responded straightaway. Dublin Zoo is part of the fabric of Irish society, but we hadn't realised just how much. We had hoped to raise €200,000 to give us a bit of breathing space, but the outpouring of generosity was beyond anything that we expected. The public gave us almost €3 million to keep going – enough to care for the animals for almost five months. And then the government stepped in, announcing that there would be €1 million shared between Dublin Zoo and Fota Wildlife Park in Cork, so that gave us another boost. We knew we could take care of all our animals until the restrictions lifted and our gates opened again.

Two weeks after I took part in that appeal, I was back in the news again. After 36 years, I was stepping down as operations manager and elephant keeper at Dublin Zoo in the new year. It was a big decision, one that wasn't taken lightly, and it wasn't a perfect exit. As the pandemic continued, it was obvious that restructuring was needed. I was called in one day and told there was going to be a complete review of roles, including the positions of senior curator and operations manager, which were similar. People knew that I had an interest in potentially setting up my own consultancy business, but it was still a shock to hear that my position was under review.

It wasn't only me. I heard from colleagues all over the world that similar management positions at my level were being axed. I had multiple friends who had either been demoted or let go. I sat down with Leona, and we decided that if there was a redundancy offer, we should see it as an opportunity to break out on my own. I had thought about it before. I was already being asked by other zoos to advise and help them with their elephants. But when you have a life-long association with one zoo, as I did, the thought of breaking away is daunting and scary.

But I did break away. I took a redundancy offer and set up Global Elephant Care. I would still be a consultant for Dublin Zoo, but there were elephants around the world that needed help. I wanted to motivate and inspire the next generation of zookeepers, and the current generation of elephant keepers.

The seeds of my departure had probably been planted during the months at Knowsley Park. On reflection, I can see that psychologically it was really beneficial for me because it took me out of the zoo environment I had known all my life. I had never been away from the zoo for a prolonged period. But I found myself in a different zoo, using a different skill set and leading a training programme. A natural detachment was taking place: the zoo was learning to do without me, but I was also learning to do without the zoo.

As Kavi and Ashoka began their new life in Sydney in December 2020, I was about to begin mine. Leaving any job, it is good to know that you have had a positive impact. I was essentially the face of the zoo for years, and it felt good to know that working with a couple of great directors along the way, I had contributed to a transformation that had made Dublin one of the finest zoos in the world. I was also confident that its new director, Christoph Schwitzer, would transform it even further, and I was further reassured when the zoo's 10-year plan was published in 2021.

I also knew the zoo's future was in the secure hands of valued colleagues I was leaving behind. Keepers like Ciaran McMahon, who had been my constant support, especially when decisions needed to be made, and deep-thinking Donal Lynch. Both had been as passionate as me about our elephant programme.

Shortly after I left, my brother James and his wife, fellow keeper Rachel O'Sullivan, also announced their departure from Dublin Zoo. They left to run Rachel's family farm on the Wicklow–Kildare border. For the first time since 1958, there was no Creighton surname on the staff rota. My sister Margaret's sons, Anthony and Eric McClure, joined as keepers in 2009 and 2010, but Anthony is now the only Dublin zookeeper in the family.

I have spent more time in the Phoenix Park than in any other place in the world. Driving into it every day added to the joy of coming to work. Every season brings a different beauty, but my heart lifted a little higher every autumn driving through the Cabra Gate into a kaleidoscope of colour. I've always felt that working in the park is good for you emotionally, spiritually and physically. A lot of the old keepers, including my father, are living to really good ages. I have a theory that their secret is the hard physical work of a keeper while breathing in the park's fresh air. It's almost a therapeutic job. You get to work with incredible animals and interact with fascinating people, in an environment that's conducive to good health.

In almost four decades, I don't remember a day where I didn't want to jump out of bed and go to work. The pulse of the zoo had been my heartbeat for a long, long time, so to leave it was always going to be a wrench. I remember driving out into the park on New Year's Eve, thinking, 'I'm not back there tomorrow, I've actually finished.' I pulled in the car and got really upset, wondering if I had done the right thing. I was really sad. I realised I wouldn't be going in the next morning to see Dina and the rest of the herd. I felt like I had betrayed them, and all the animals. These were the animals that I had constantly battled for. I had always looked for better conditions and better resources for them, and now I was turning my back on them. I felt an enormous weight of overwhelming guilt. But then I went home, where Leona reassured me that I was doing the right thing. I knew I had to make it work.

What's funny is that, while I've physically departed, I'm still constantly dreaming about the zoo, and it will often be of older times, of the old Bear House or the old Lion House. They are good dreams: sometimes they're about things that really happened, and sometimes they are new scenarios and situations. The other night I dreamed that the old chimps had escaped. I found them in the old black bear pit, and I was trying to decide how I could get them back to where they were supposed to be. It was so vivid and so real. I woke up and said to Leona: 'I'm dreaming about the f***ing zoo again!'

Sissy

I worked with an elephant in the US called Sissy. Captured in Thailand as a very young calf, she was taken from her family and first displayed as Sis Flagg at a petting zoo in the Six Flags Over Texas amusement park. When she became too big, she was sold to a zoo and renamed Sissy. During a storm and deadly flood in 1981, she was one of many animals swept away. Thirty-six hours later, when floodwaters began to recede, her trunk was spotted wrapped around a tree limb. Only her trunk was above water, allowing her to breathe. Her incredible survival briefly made her a media darling.

Over the next three decades, Sissy moved between different Texas zoos, confined to small enclosures, held down with straps and chains, and regularly beaten with baseball bats and axe handles. She didn't respond well to other elephants, and often displayed aggression. When a zookeeper was found dead in her enclosure, although it was unclear what had happened, she was labelled a killer – a mad elephant.

She made the news again in 1999 when a video emerged of her being beaten into submission by keepers at El Paso Zoo. Its then director, who later resigned, defended the action as 'discipline', but the zoo was charged with violating the Animal Welfare Act, and the local community voted for Sissy to be moved to the Elephant Sanctuary in Tennessee.

Sissy embodies so much of the badness and the consequences for elephants in human care. Taken from her family as a calf, she had no time to learn elephant skills and socialisation. She was trained with dominance and pain, used as a prop, confined and restrained. It is no wonder that she showed signs of aggression towards keepers and did not relate well with other elephants. Why are we surprised when elephants react like this after years of abuse, after years of the hook? Why are we surprised when they defend themselves against violence with violence?

Sissy endured trauma after trauma, but she has found a new life at the Tennessee sanctuary, where she wanders happily alongside other elephants. She is an incredible elephant despite her age and considerable damage to her trunk. She eventually overcame her fear of water and can be seen splashing and swimming. The move also gave her the confidence to lie down outside for the first time in years. However, her fear of people and confined spaces means she is still unable to lie down in her inside house.

Kristy Sands Eaker invited me out to Tennessee to consult about some of their elephants. We are coming up with a programme to make Sissy lie down. She's very nervous of the sand, so I told the keepers to start with a little bucket of sand

and put some food around that. She has to move a little sand to get the food, and this has started creating positive associations. Gradually, the sand pile will be made bigger and bigger until she has to step over it and feel it under her belly. In time, we hope that she will put her belly on it, or lie down on it, and realise that sand can be supportive, that it can feel good. Then a sand pillow can be built that she can interact with, and she will start to associate her indoor space with comfort, positivity and feeding.

But it's a huge psychological barrier for her to overcome. It is a testament to how intelligent these animals are, and the injustices that have been done to her, that Sissy associates going into a small space with bad things, with violence. She is trying to protect herself.

Humans need to stop creating Sissys. We need rules and laws that will prohibit free contact, the hook and human violence against elephants everywhere.

TWENTY-ONE

THE ELEPHANT MAN GOES GLOBAL

A ny worries I had about leaving the zoo were quickly dispelled. The phone didn't stop ringing with requests to travel all over the world to assist with elephant training programmes. They all wanted to do it the Dublin way.

At the invitation of Thane Maynard, the Cincinnati Zoo Director, who has been a great friend and adviser, I worked for almost five years with the Cincinnati Zoo & Botanical Gardens in the US, which had a very old elephant house. Since 1993, the zoo has invited naturalists and scientists to speak on wildlife issues and global conservation efforts as part of its annual Barrows Conservation Lecture Series. In 2018 I was invited to talk. My lecture was: 'Giant footsteps: The future of the Asian elephants in human care'. When the zoo decided to build a large new habitat on a couple of acres of reclaimed land, I helped raise millions of dollars for its 'More Home to Roam' project. I'm very proud that four of Dublin's elephants will travel to Cincinnati Zoo at the end of this year to become its first multigenerational herd. I was instrumental in recommending to the EEP that this is where one family grouping from Dublin that is naturally separating from the other should go. It is very exciting to think that what we began in 2006 in Dublin will now be replicated in Cincinnati, and I am looking forward to watching the family develop over the next decade.

There are not many breeding herds in North America. When mother and daughter Yasmin and Anak arrive with their sons Kabir and Sanjay, it will resonate throughout American zoos to see a family herd. The cows will breed with a bull, bringing a whole new bloodline to that continent, and it will have a ripple effect when other zoos see their birthing process. Unfortunately, many zoos still try to manage births, even tying up or restraining cows in labour.

A matriarch and her daughter (like Yasmin and Anak) always stay together, and they are the key components of elephant society. Yasmin will be leaving her sister Dina behind, but in the wild, a younger sister and her daughter would go off like this and form the nucleus of their own multigenerational herd. That's the unbreakable bond, and in a lot of cases the daughter of the matriarch will eventually lead the herd – it's almost like a hereditary title.

We should always aspire to having a multigenerational herd of elephants. That means a family that lives, works and plays together. For too many years, we've put unrelated elephants together in zoos. They've been tolerated but they have never thrived because they were not part of a herd. Putting elephants together in one space is not the same as building a multigenerational family herd. Sharing space and having a cohesive meaningful relationship is very different. However, there is still a lot we can do for elephants that are together but don't necessarily get on, like giving them room to be apart or putting in physical barriers like fallen trees and sand piles.

There are now recommendations that every elephant should have company. Some organisations, like the EAZA, say that there should be a minimum of three or four elephants together, but in my experience, these have to be related elephants when living in the same habitat. I've seen all hell break loose when elephants with circus backgrounds, which don't allow for normal elephant behaviours, are put together. For some single elephants, keeping them busy and active may be a better solution.

I've also been advising the remarkable Elephant Sanctuary in Tennessee. There's a wind of change in the US, with more zoos admitting that they made mistakes with how they managed elephants, and looking to the sanctuary as a potential new home. They're amazing people doing an amazing job with elephants that arrive bankrupt, emotionally and physically, because of being in human care. These animals have spent their lives in circuses, chained up or in zoos with poor conditions. The result of human dominance and lack of care is elephants who are sad, or who have major behavioural issues. In Tennessee, they are being given new lives. It's wonderful to see them starting to show normal behaviour. They can lay down on the grass outside, swim in a lake or river, push over a tree or interact with other elephants or the white-tailed deer that roam this vast area. The future for elephants is in these bigger spaces because it gives so many choices.

I'm helping them design for bulls and plan for the future, so they can take in as many elephants as possible.

In Australia, I've been working with Lucy Truelson, Ben Gee, Erin Gardiner and the team at Melbourne Zoo, which has a very successful elephant programme, including the births of three calves within weeks of each other at the end of 2022. Like most city zoos, space is restricted, and they are moving their elephants to a 24-hectare site in Werribee Open Range Zoo in 2024. Werribee's director Mark Pilgrim was the director of Chester Zoo and a driving force behind its elephant programme. I've been helping with the design of the new habitat, which is going to be a game changer internationally – not just because of the space the elephants will have, but because of the opportunity to create different topographies and landscapes for the herd, from savanna-type grasslands to forests with pools, overhead feeding and different temperature zones … all the opportunities that a wild elephant would encounter during the day. This project aims to mimic natural feeding and social patterns.

One of the things we need to do for elephants, and other animals in human care, is to enhance their habitats without interfering with their social dynamics. Currently, a herd might be feeding or interacting, and keepers have to bring them indoors to replenish an outside area. Our intentions might be good, but they may still disturb the social and intimate moments of herd life. A bigger, more spacious, habitat means that humans can become even more irrelevant. At certain stages during the day, hay nets can drop or browse can be put in areas that the herd migrates through.

In the wild, the calf spends most of its time looking at its mother's backside, because they are constantly moving. However, in a lot of zoos, you will see them going around in circles since they don't have a lot to do. Elephants need complexity of movement. They move for resources: where they know there's a waterhole or fresh browse. The Melbourne project excites me because we finally have enough space to work with. The elephants will have a house, but they will always have access to the outdoors, where there will be a massive input of resources. It's as close to the wild as we're going to get and close to the wild means wellness. I think once the world sees this, it will require every zoo to check its conscience and compare the conditions of its elephants with those at Werribee.

In November 2022, I found myself on a long-haul flight, on my way to spend a month at Al Ain Zoo in the United Arab Emirates (UAE). When Al Ain and Sharjah Safari announced that wild elephants were being bought from Namibia, it caused a huge controversy within the international zoo community. The sale, which Namibia claimed was needed to reduce elephant–human conflicts, triggered international condemnation by animal advocates, scientists and governments, and resulted in Al Ain's expulsion from the EAZA.

Al Ain took eight wild elephants from the wild and brought them into a zoo situation, something that should never have happened. I went there because the animals needed help. I had no part in the process of bringing the elephants to the UAE, but I decided I would be part of the process to ensure that they are being fed and cared for appropriately and that they are trained for veterinary access. I only agreed to travel to the UAE after consulting with my wife, family and other people, including members of the EAZA Elephant TAG. I knew I had to stay away from politics (politics wouldn't help these elephants), but

I wanted to ensure that life plans were being put in place to care for them.

On the 130km drive from Abu Dhabi to Al Ain city, the magnitude of the task started to sink in. Usually, when I visit a zoo to give elephant advice, I'm going into accomplished established teams that only need fine-tuning. I might just be there as a consultant on foot complaints, injuries and trauma, or some issue with a facility's design. On this job, I was beginning with a blank canvas. Neither the elephants nor the keepers had ever been trained.

In the hotel, I looked over my presentations and practised slowing down my accent. I'm a fast talker, even by Irish standards, and the cultural differences between an Irishman and the people who live in the UAE are significant. I needed to make myself understood. I quickly found that videos of work and training I had done over the years were an effective way to overcome language differences.

On the first morning, I considered turning on my heels and heading home. I had immediate concerns about the facility's design and the safety of both elephants and keepers. The elephants had done lots of damage to their habitat. They had broken electric gates, and scraped plasterboard from a poorly designed exterior wall, exposing steel pins and insulation that could be dangerous if eaten. There were safety issues with the design of the protected-contact wall, with drinking sources, with hanging cables … the list was long!

However, when I met the elephants, I knew I had to overcome the design flaws and make it work. All eight looked very well physically, but they were vulnerable and confused. They were not happy for me to come close, even behind a protective barrier. Ears flapped out and trunks were curled under, which is threatening elephant behaviour. They charged at the fence. I quickly spotted the group's matriarch, Zulu, and began to identify each individual, noting characteristics like ear creases, tears and tusk definition.

The keepers were an energetic team of young men, who were keen to learn from the start. An international group picked from around the zoo, they had never had any elephant experience. They came from the UAE, Afghanistan, India, Jordan, Pakistan and the Philippines, and now I was a Paddy adding to the mix. I would continue to be impressed by how much they wanted to learn.

They had been herding the elephants like cattle, so I had to quickly defuse any previous management methods and get the elephants into a calmer, quieter space. In an attempt to bond with the elephants, they had also been trunk feeding (placing food in the elephants' trunks). Not only had this confused the elephants, but it was also potentially lethal as an elephant could suddenly lash out with its trunk and hit someone.

Aside from the elephants' welfare, what made me decide to stay was the zoo's willingness to accept that urgent changes were needed, and its commitment to resourcing those changes. I told them the international community was looking at them, and that they would have to show they would give these elephants a meaningful life.

Aside from the challenges I could see, there was one I could feel: the heat. I remember going to Ibiza for the first time (I think I was 17), getting off the plane and feeling like I had walked into a warm room. Here you could multiply that feeling by 100. The locals told me – the red-headed Irish man bursting with heat – how lucky I was to have come at the coolest time of year, and it was 37°C! It gets even hotter there because it is the desert, and I was told temperatures can reach up to 50°C in the summertime. I could barely spend more than five minutes outdoors, and I knew this would impact the elephants, too, even though they were African elephants.

It was also a challenging, rocky landscape. There were mountains in the distance, but few signs of colour or life, just the occasional roller bird, which is a beautiful green and turquoise. It was the polar opposite of what I was used to in Ireland. On that first day, a TV ad from years ago came into my head. It was the one for Harp beer, where a man in the desert, with the sun beating down on his face, says, 'You could fry an egg here if you had an egg; and you could sink a pint of Harp if you had a pint of Harp.' Like him, I felt out of my comfort zone.

In very hot countries like the UAE, it's important to figure out the shadiest areas for elephants. Food can be buried, and root balls or browse scattered, to allow an elephant to choose whether to stand in cooler areas and feed or spend time in the hot sun. The environment also contained other hostile elements. One day, as I lifted a hay bale, one of the keepers pointed out a saw-scaled viper sliding away. Not only is it one of the fastest-moving snakes in the world, but it is also highly venomous, and a human, or animal, could die pretty quickly from a bite. The elephants recognised the danger, too. When they subsequently encountered one, they killed it quickly to protect the herd.

I recommended distributing small piles of pellets all over the ground of the inside habitat to increase movement and prolong feeding. However, African elephants do not usually feed from the ground – they are elevated feeders, and in the wild, they will stretch on two back legs up into the trees. When I arrived at Al

Ain, the herd was being fed twice a day by dropping food. They finished their morning rations outside after just 40 minutes. There were indoor hay nets, but the holes were too large, large enough for an elephant to push its trunk through and potentially get entangled. Hay nets for elephants should have gaps smaller than the width of a trunk so that the elephants have to work hard to extract the hay. Adding pellets and branches into the hay nets also makes them more interesting to interact with, and the bending and stretching that is required will strengthen their neck, shoulder and trunk muscles.

I asked for pieces of browse of varying weight and size. Even without leaves, African elephants will spend hours removing the bark from browse, which is very beneficial for tooth health and trunk fitness. This was a great way of keeping the elephants mentally and physically occupied when indoors. We began planning a broader range of feeding options, both inside and out. The way food is prepared and presented is key for the overall physical and psychological wellness of an elephant herd. African elephants, in particular, are tenacious and agile browsers in their natural habitat, and they spend a considerable proportion of their waking hours searching for and eating food. I began to spot areas where we could introduce hoists, feed boxes, tubes and pipes at different levels in order to challenge the elephants, both physically and mentally. We need to keep them busy, and social feeding also helps the development of positive long-term relationships within the herd.

The core of any elephant programme is safety, for both the elephants and the keepers. I needed to ensure that all 12 keepers would go home safe at night. The house was massive, and the outside yards were substantial, but to figure out the weaknesses, I stood inside the protected-contact wall where the elephant stands, and then where the keepers would be. I worked out safer ways to move the elephants between their indoor and outdoor areas and to allow training.

The space that would be allocated to the herd in the safari park was considerable. But I could see areas where elephants might potentially injure themselves on rock formations. It was great to have such a large area, but I explained that you can let elephants out into a 40-acre site, but if there's nothing there, they will just return to their house. Their habitat needs to be packed with resources and choices.

What I preach, everywhere I go, is that elephants need care 24 hours a day. Asian and African elephants have different needs, but their care has to be activity-based, fitness-based and nutrition-based because they just don't like standing around. Proper care also requires profiling individual elephants to determine what it is going to need throughout its life, as a calf, a toddler, an adolescent, an adult and a geriatric. An elephant needs a life plan, and the resources to support it.

The herd was given extended access to the house and outside sand areas, and I suggested adding access to the outside area at night, particularly when it's cooler. I recommended that large logs, root balls, scratch rocks and walls, plus a designated mud wallow area, be introduced to offer the elephants a good choice of activities.

The herd loved caking themselves in mud in the wallow we built. The skin of African elephants is far more wrinkled and creased than that of Asian elephants, and in many ways, wallows are more important to them than pools. As the mud and water get trapped in these creases and folds, it helps them regulate their body temperature, cooling them down, and also protecting their skin from the sun and insect bites. In addition, the whole activity is highly social, so bonds are reinforced and play behaviour is instigated. The positive effect is continued with the scratching and stretching behaviour that usually follows, contributing to sensory pleasure and skin maintenance. The whole activity is physically strenuous, which helps to strengthen muscles and burn energy. After activities like this, elephants tend to enjoy a long period of calm and rest. I asked for a truckload of mud/clay to be added at least every week, and that the wallow be replenished with digging and a hose every day. Small pools were created to help a new calf have her own playtime.

Even I was surprised by how quickly the herd responded to positive reinforcement training. As soon as we started interacting with them, I saw glimmers of curiosity. We began by giving keepers feed bags, asking them to call the elephants over and feeding them on the ground. Next, I asked keepers to hold and move a target stick from one side to another while feeding the elephants, and to lean it towards the elephant when throwing the food rewards. This allowed the keepers to practise holding the target sticks while feeding, and it desensitised the elephants to the presence and movement of the sticks. After four brief sessions, on day two, three of the elephants were happily feeding with a target stick against their forehead.

By day three, we were slowly building trust with them all, but the elephants still let us know that they are wild by running at the fence line. I kept reminding the keepers that they should never make the mistake of viewing an elephant as a friend, even when trained. 'They're using you for the reward, they're using you to get what you have in the bucket, not because they love you.' Keepers need to understand this because safety has to be paramount. They will always need to be focused, to look out for each other. You need to be at your best mentally and physically when you're working with elephants. I repeatedly told them that if they had any issue – family, health, whatever – they were to put their hand up and say, 'I don't want to train today.'

For a month, I worked between 10 and 12 hours a day, including weekends, but it was worth it. We made huge progress, and the elephants exceeded my expectations with their willingness to participate and their ability to learn. They responded with kindness and structure. Their keepers also soaked up new knowledge and mastered new skills. Further modifications and safety measures were agreed upon so that training could progress.

Before leaving, I trained the herd to respond to a bell. When I ring the bell, they know to come quickly because there will be food waiting for them. This device can also be used when they move to the safari park if there's a need to recall them or an emergency.

I returned to Dublin far more hopeful. There is a long way to go before we can make any claims about elephant wellness, or welfare, in the UAE being up to international standards. I know there are huge challenges ahead, but there is a commitment from Al Ain management to reach those standards and to create a world-class habitat and a world-class elephant programme. My work with this herd also has the potential to make a huge difference for elephants in the entire region. That was always my ambition. I've already had approaches from a Dubai safari and another zoo in the Emirates region.

While in the UAE, I remained in contact with my other projects, including the most exciting: the return of 13 African elephants born in zoos back to the wild. The elephants were located in an 8-acre habitat at Howletts Wild Animal Park in Kent, which has one of the most successful breeding herds of African elephants in Europe.

And before the year ended, I was back there.

This is the first time that a herd of elephants has ever been rewilded anywhere in the world. A lot of zoos have talked about the possibility of doing this for a long time, but Howletts is actually going ahead with it. It is a brave plan, and not without risks, but nothing in life is. Howletts has already introduced gorillas back to the wild, and it has had losses, but there have also been successful reintroductions. There is a great chance that this could work.

Howletts and The Aspinall Foundation are doing this alongside the Sheldrick Wildlife Trust, which operates an orphan elephant rescue and wildlife rehabilitation programme in Kenya. My role has been to ensure that the animals are physically and psychologically prepared for that journey, which they will make by plane and road in crates. When we began at the end of 2021, none of the elephants had ever been target trained, and aside from team leader Natalie Boyd (an excellent elephant keeper I've known for a long time), the keepers had very little experience. However, Courtney Mansfield, the assistant team leader, and the rest of the team put their heart and soul into making this work, and what we achieved is remarkable. The elephants come into crates together, desensitised to the gates closing behind them, and stand happily feeding and interacting.

The elephants will go into a 50-acre reserve in Kenya, Shimba Hills, that will be managed and monitored for a long time, with gradual dietary and environmental changes. Then there will be a selected release. There are no lions or predatory animals in this particular spot at the moment, nor poaching; the odds are stacked in the elephants' favour. There's a lot of work happening on the ground over there to prepare, including the eradication of the tsetse fly, which can be quite dangerous.

The transfer to Africa from Howletts will happen at the end of 2023. To know that these 13 elephants will re-occupy part of their ancestral homeland and walk in the footsteps of their ancestors is thrilling. To have had a small but significant part in this plan to take elephants who have lived in human care back to the wild is such a privilege. It is huge for conservation, and it has the potential to shake the conscience of the international zoo community. Other zoos may be inspired to do it too. I am not suggesting that all elephants can be released, but we can learn from the successes or failures of this project.

I feel like I'm only peaking. I've had opportunities to accumulate learning throughout my life, and I've so much more to teach and learn. The Namibian elephants at Al Ain, Howletts, Melbourne Zoo and the Elephant Sanctuary in Tennessee are probably four of the most important things that are currently happening with elephants around the world, and I'm extremely honoured to have been asked to consult on them all. The past few years have had their challenges, and personal consequences, because of the amount of time I've been away from my home and family. However, I'm hoping that they'll be able to travel with me more, and I realise I will need to employ some other like-minded people to help. I realise now I can't be everywhere.

I'm hoping to look back in a decade's time and say Global Elephant Care made a difference. I want to take what I have learned and influence the lives of elephants for the better. They need people batting for them. They need a voice, and I want to talk on their behalf.

Behati

My WhatsApp and text chats are always packed with elephant images, but in February 2023 I scrutinised the pictures and videos I was being sent from the team at Al Ain Zoo with even more attention than usual.

During my first visit to Al Ain, I noticed that Masego, one of the mature females (probably the second-oldest female) looked quite pregnant. I could even see movement around her abdomen. Unfortunately, at that point, the elephants weren't trained well enough to be able to take blood samples to confirm this. But I told the care team that all the indications pointed to her being pregnant.

I instructed the team on what behaviours to watch for in Masego, and I told them to pay attention to how the other elephants, particularly the young females, behaved around her.

By February 2023, more physical changes had become apparent: Masego's mammary glands were getting bigger, and she was becoming a little bit more lethargic, which was very evident from the videos that the team were sending me. I advised the team to watch out for a swelling above the tail as the calf got into position for birth, to be calm and quiet near her, and to ensure that she, and the rest of the herd, had plenty of food. I also recommended that they keep all the food in the sand area of the habitat, so she would give birth on sand.

And a beautiful, healthy female calf was born. It was an amazing moment to get the video of the new arrival. I was relieved to hear that Masego was allowing her calf to suckle, but it soon became clear that she was not allowing it to lie down because she was worried about predators. She was using her trunk and feet to get the calf to stand up and walk. I explained to the team that she was not going to rest, or let her calf rest, until she could sense there was no one nearby. She wanted to protect her calf, just as she had learned from her years in the wild. Her instincts told her that there might be lions or something else that might harm her calf, in the area. I advised that everyone stay away from the herd for as long as possible, and only go near to replenish food stocks.

It was fascinating to see how the birth changed the whole behaviour of the herd. The elephants became focused on this wonderful new addition and on bringing her into the herd. They gathered around her in a defensive circle to let her sleep safely. Even the young females, who were not even at breeding age, were lifting their front legs, trying to encourage the calf to suckle. They, too, were learning for the future.

An elephant birth is always wonderful. In the zoo, a new life was always an incredible form of stimulation – both for the animals and the keepers. Being witnesses to birth and new life, whether a primate or a cub, is such a pleasurable part of what we do.

I got to the UAE as quickly as possible. I'd been guiding the team online and through video messaging, but I wanted to see this new infant and check the behaviour of the herd to make sure that all members and the calf were getting the right opportunities to succeed.

The new arrival was given the name Behati. Researched and chosen by my daughter Mia, the name comes from Africa and means 'blessed, she who brings happiness'. And she certainly did bring a lot of happiness to the herd: to see them all working together, as a family unit, to protect this calf was inspiring.

She was a beautiful calf. African elephant calves have a very unusual, but cute appearance, because of their very large ears and wrinkled bodies. I spent my birthday with her and the herd. Masego constantly used her trunk to guide her calf away from me, but she soon felt secure enough to join in training, while keeping her calf close by her side.

Watching Behati with Masego was a real wake-up moment for me. This little elephant was conceived in the wild, in the grasslands of Namibia, and for reasons that she, and her herd, had no control over, she now faces a life in human care. It really hit home to me how important it is that I get this right – for her and for the elephant care team. I realise all their futures depend on the choices humans are about to make for them.

I sat down with the Al Ain management team and the keepers to ensure they were aware of their responsibilities towards this elephant and her herd. They need to show the international zoo community that there is a life plan for little Behati and the other elephants. She could potentially spend 50 or 60 years in human care, and we have to get it right. We have to be aware of the different requirements of her different life stages – from this time as a playful young calf that's adored within the herd, to being an elderly elephant – and be prepared for every eventuality. She is a beacon of hope for her species, but we have to show that despite being born in a zoo, every aspect of her care allows her to reach her full potential.

It's a huge challenge. However, senior management at Al Ain were quick to show their commitment, giving the necessary resources and making habitat enhancements. A very large safari area is in the works. It's a significant space where the elephants will have mud wallows, overhead feeding and a vast landscape of different topography changes, to ensure freedom of movement and expression. Any gaps are now being filled to allow a whole biological repertoire that is comparable with wild elephants.

Al Ain management has also voiced its commitment to a comprehensive conservation programme, and we are looking at various ways it can contribute to conservation in the wild, in the elephants' homeland of Namibia.

The next three years at Al Ain will be very exciting for me as a consultant because I am determined to offer this little elephant every opportunity she needs to succeed and have a meaningful elephant life. And hopefully, in the future, she can make a valuable contribution to elephant conservation, not only in the UAE but further afield in that region. And who knows – maybe even globally!

TWENTY-TWO

WELFARE BEGINS
AT HOME

When I was growing up in Blanchardstown, people would often come up to the house with animals that had been injured because they knew I would take care of them. But we weren't the only people doing that. Jenny Powell was a very well-spoken English woman who lived for years surrounded by animals in a little ramshackle caravan, without water or heat, near Mulhuddart. You would see her outside the shopping centre, and she might have five goats and four dogs following her on leads.

Jenny loved animals, but she was eccentric, and people were cruel to her, particularly kids. She used to get terrible abuse going through the estates, and she would often call in to ask me to walk her through: 'Gerry, would you walk me up a little piece because the bullies will take my food and pick on my animals.' I was known from the boxing, so I would walk with her and say to the young fellas around Corduff: 'I will batter you if any of you touch her or the dogs.' I wanted people to associate her with me as her friend. It meant that when I wasn't there, because of the respect I had in the neighbourhood from the boxing, they would leave her alone. It got to the point when Jenny would be walking past young fellas telling them, 'If you say anything to me, Gerry Creighton will be up here after you.' It was funny. After a while, they accepted it and left her alone, but I used to have to walk her up regularly.

One day I went up to her caravan with a bag of dog food. When she called me in, there were two goats asleep on each side of her, plus a chaotic mix of dogs, cats, horses, mice and rats. They were in perfect harmony. It just shows you the power of the person who brought it together.

Another time my doorbell rang, and Jenny was outside with a big old pram, the kind that had bouncy shock absorbers. In the middle of the pram was a giant rat, gasping. I said, 'Jenny, the poor thing is after being poisoned, it's about to die, and it's not good to be handling it.' She insisted: 'Gerry, if anybody can do anything for it, you can,' and she was holding up this giant wild rat. I kept telling her to put it down because she might get leptospirosis from its urine, but she must have been immune. I took the rat and put it out of its misery.

She would regularly call the zoo looking for advice from me, or Da when he was there, on how to treat an animal. 'Hi Gerry, I have a goat up here. This is the problem...' Or she might say, 'Gerry, I was just thinking, I have a bit of space up the back if you need me to take any of the giraffes or any of the big animals.'

She was a fascinating character. I think she had lived in Australia and Canada. She told me she had trained horses at one stage, and worked in a circus or carnival, but something very bad had happened to her. In my heyday, when I was going to nightclubs, I would see her sitting outside Brown Thomas on Grafton Street reading fortunes for people, and she'd always have a few animals with her.

Jenny and I were two people from very different situations and backgrounds, but our love of animals brought us together. She had the appearance of somebody who was down and out (and I don't mean that with any disrespect), and she didn't have washing facilities, but I loved listening to her stories about working with Arabian horses. She was a very intelligent, smart lady, and I had incredible admiration for her. At the heart of everything she did was her desire to care for animals – every animal.

When I left the zoo, one of the first jobs I took on was with Dublin City Council, to help with their dog welfare unit: to do some training with animal welfare expert Sabrina Brando and to support their people when dealing with dangerous dogs. They knew my skills with big cats and other animals would be very useful, and I could immobilise with a tranquiliser dart gun if needed.

There have been lots of issues in Ireland with families getting dogs during the pandemic lockdown, and not thinking about what would happen when people returned to work and school. Puppy farming is also an ongoing concern. However, in Dublin, there has also been an influx of particular breeds that are a growing welfare problem. Breeds like bullies, cane corso and different types of mastiffs are being used as fashion accessories. Criminal and drug gangs are even putting money into breeding these dogs because they are selling for thousands. Bullies are a cross between a bully breed (like a pit bull) and a Patterdale terrier, which is supposed to give the dog a calm disposition. They come in stem cells and the females are artificially inseminated. Young men buy them as you would buy a shirt – in XL. Indeed, that's how they are marketed: in miniature, standard XL and XXL.

The young men that buy them crop their ears (a mutilation of a beautiful animal) to make them look more intimidating. And because the cutting of a dog's ears is illegal in Ireland, the dogs are confiscated. On one occasion, I went to a pound, and it was housing about 30 of these XL bullies. Ninety-nine per cent of these dogs would kill you with a kiss. They are so sweet, I would have my kids up walking them with me. However, some of them have been turned aggressive because of human domination or because this is what they have been trained for, to be used in criminal and drug activity.

I have had quite a lot of call-outs, often to Garda stations, because of these dogs. One call-out was to deal with a ferocious dog called El Chapo, which had attacked a garda during a raid, destroying his leg. I had to immobilise him with a dart gun, because he had gone beyond the point where you could put a lead on him, and he was then put into a crate and taken to a secure area in a shelter. Another time a cane corso-type dog attacked a child. I immobilised the dog, so the police could get him up to the pound.

It is a real issue when people don't understand the consequences of keeping dogs as fashion items. During my animal welfare work, I have seen a lot of Belgian malinois or Belgian shepherds. These dogs have incredible intelligence, and they are phenomenal athletes: they remind me of monkeys, as they can run up a straight wall for five or six metres. You would often see their trainer bending at the back, and the dog uses their back to propel themselves four or five metres over a wall. The Belgian Police Force uses them, and they are often used as guard dogs. However, people are buying this powerhouse of energy and intelligence without thinking about what it entails.

One Belgian Malamute I had to immobilise had been surrendered by his owner to the shelter. He was in a frenzy in a small crate. They couldn't calm him down, so I had to sedate him with the dart gun. And because he had been surrendered (he had attacked his owner, giving him significant bites), he had to be euthanised. This went against everything I have ever stood for: I was devastated knowing that this magnificent animal, in perfect health, would be put to sleep forever. Then, as if to reinforce the point, as I drove home I heard a piece on the radio about how 30 dog breeds had been tested for intelligence and agility, and the Belgian Malamute had come out on top. This is the issue. People buy these amazing dogs (for considerable amounts of money) and think that it's appropriate to keep them in a tiny back garden or a small run, just bringing them out for feeding or a walk. But these are animals that need to use their advanced intelligence, thought processing and boundless energy. They need to be working, to be useful. Humans need to use these animals when they can do the most good.

I have seen a couple who are hyper-aggressive. These are not bad dogs, but they are brought into situations where there is no thought process. There is no plan. This reminds me of zoos years ago, when all these animals were brought in with no discussion about how to deal with their physiology, their psychology and their biology.

One Malamute I saw in a shelter was doing constant figure

eights. His head was now permanently turned to one side, and one of its legs was starting to wear down from the constant repetitive movement. That is called stereotypical behaviour. You can see it in circus elephants, even some zoo elephants, when they push a leg from side to side. It is the same motion of an animal walking but in a standing position. It is almost as if the body wants to feel like it is moving, using up energy, but it is actually a coping mechanism, and it is very hard to reverse it.

I get call-outs from animal welfare groups throughout Ireland, including the DSPCA or My Lovely Horse Rescue. It is good to be able to support them because sometimes these animals can be emaciated, hungry and fearful. I once spent a week in Clondalkin trying to catch a horse. The rope had grown through his face, and it was all infested with maggots. We got him treated, and they ended up calling him Gerry. There are hundreds of urban horses up in Dunsink, and I spent days there in a wet ditch waiting for a particular horse to walk by. The horse had been caught in a snare, and the wire had eaten through his leg. Another poor horse we helped was extremely stressed. She had been put in a head collar, and it had been left to grow into the skin at the back of her mane.

I have gone into houses and seen dogs in horrific conditions. In one place, we took out 13 or 14 dogs, all different breeds but all destined for the pet trade. People are just mass-producing pups, and the animal's welfare is never considered. On another occasion, I was called to Northern Ireland where a beagle had broken away from a hunting pack. Farmers were going to shoot and kill it because they were worried about their sheep. Eventually, we got the dog under control. If anyone calls me to help an animal, I'll go: I cannot turn away from an animal in need.

I found the *Tiger King* TV series extremely distressing because it showed the abuse of animals – cubs being pulled away from their mothers to be hand-raised so that humans could be with them for their own egos. It is a huge industry, and I find it abhorrent that animals are still being bred for entertainment and to be kept privately. In the US, there have even been tigers found in apartment blocks. There are welfare laws that prevent particular species from being exchanged, but hybrids or mixed-breed tigers are not governed by these legislations. You can go to farms in the US and pay $500 for a lion or a tiger. People buy them as cubs, ignoring the fact that they are carnivores and need a certain lifestyle, and they certainly shouldn't mix with humans. Nobody is thinking about what the animal's needs will be over its life.

I will help anyone who wants to improve the lives of the animals in their care. I was contacted last November by Sabrina Brando, who was working with a small zoo in Bordeaux, France. They didn't have much money but were trying to do their best to improve conditions for their jaguars, so I offered my services for free to help change and adapt the habitat. It was important for me to do that, and it was life-changing for the jaguars. For the first time, they had the opportunity to climb, to do what their biology is designed to do. We created tiers with tree trunks. We banded them together into a three-dimensional frame around the habitat, so they could climb from the ground up to high vantage points, with views across the zoo and the stimulation of seeing prey species (sheep and goats) in the distance.

One of the things I miss about working at Dublin Zoo is talking to visiting kids. The zoo's Discovery and Learning Department reaches approximately 60,000 schoolchildren and learners every year. If I have inspired just one of them, then it has been a job well done. Anyone who keeps an animal should understand its needs and educate themselves about its welfare, and we need to start with children, so they understand the time and investment involved – whether you decide to buy a hamster, a pigeon, a dog or a cat.

We all know the benefits pets can bring. Pets are most valuable and beneficial in both the early stages and the late stages of life. It's a great discipline for children to understand a species, understand how to feed it and keep it clean. Then they make the best companions when you're older when people tend to be more on their own. A dog will sit on your lap with uncompromised loyalty. It gives you love regardless of how you look, how you dress or how you smell. Its devotion doesn't change.

However, anybody considering a pet – no matter what the pet is – should first sit down with a piece of paper and write a life plan. Have a life plan, and you will be a successful animal carer. Zoos need a plan for every animal, and so will you if you are going to buy a pet. What happens when you go on holiday or when the weather is hot or cold or it gets old or sick? You need to know that you will be able to look after this animal 365 days a year, and that's no different to a zoo.

The puppy moments are lovely, but the puppy moments become big-dog moments, and then they become geriatric-dog moments. You've got to be on board with a life plan for your animal. You've got to be in for the long haul.

TWENTY-THREE

A MANIFESTO FOR THE FUTURE

The world needs a wake-up call. We have messed up hugely with elephants in human care, and it is now time to make things right. We need a plan; we need to work hard to conserve and support this amazing species; and we also need to look at and learn from them. Elephants can teach us so much. They exist in a highly complex society, and how they care for each other, form relationships and teach the next generation have lessons for us as humans.

When you study elephants and other animals that live in a complex society, like apes, you realise that everyone has a role and everyone understands their purpose. Animals live in the present. There are no agendas, and everything is done for a meaningful purpose: when they fight, they fight for territory or resources, dominance or breeding rights.

Elephants have thought out strategies for resources, movement and interaction. They have taught me to think things through, and I've learned to ask what I want my action (and my outcome) to be, and how to implement it. I now know the importance of slowing things down and not reacting without understanding the consequence, or the value to me or the people around me. It's a simple lesson to learn from them.

Elephants have also made me think about how I behave towards people, and what my expectations should be of how I am treated. One of the main things that I've learned from them is their compassion. I would encourage people to reflect and look at their society, at the bonds they form. The herd is all about investing in the next generation. They take the time to educate the younger herd members so they will be successful in terms of how they behave and interact, in terms of social play.

There is so much we need to do for these wonderful animals. We need to condemn and challenge what is happening in their native home-range countries: the cruelty, violence and misery that are still rained down upon them in the name of culture or entertainment. But we also need to try and change minds and support communities as they change. We need to dispel the ignorance that covets what elephants (and other species) have, whether for medicines, aphrodisiacs, ornaments or trophies. What gave me some hope was meeting young people in China who understand that rhino horn is made of keratin (the same substance as the human nail) and doesn't have any healing or aphrodisiac properties. And that the ivory of an elephant belongs on the elephant.

We need there to be an acceptance that elephants are no

different from us in needing purpose, in the pain they feel and in how they celebrate new life. The combination of emotion, love and excitement in a herd when a new calf is born is the same as a baby coming into an extended human family.

I'd love to see a day when we don't need zoos, but that is not going to happen anytime soon. However, zoos play a critical role in the conservation of elephants and other species. For elephants, there is no such thing as the wild. India is replacing China as the most populous country, and there is a constant human–elephant conflict where both people and animals are being killed. The race for space with humans gets faster daily, and it is not just elephants who are being squeezed out of the wild. The way parks in Africa are managed, tourist jeeps drive past the lions, and gorillas up in the mountains catch colds and flus because of the human domination of the landscape. Animals' habitats are being eroded on a daily basis. They are being extinguished and exterminated, sometimes because they put human lives and livelihoods at risk, but too often simply because of human greed.

We can't change what was done in zoos in the past. It was part of an evolutionary process and we can learn from this history to avoid repeating it. The role of the modern zoo is critical, but it must evolve. Dublin Zoo bears no resemblance to the zoo where I began work as a teenager, and it continues to change and adapt. New director Christoph Schwitzer has placed an even greater focus on conservation and education, and in 2022 he released an ambitious 10-year conservation plan for species at home in Ireland and in the wild. The zoo has a very exciting future under his leadership.

Most of the other zoos that inspire me are run by inspirational people, those at the forefront of promoting the welfare of animals in human care. I'm talking about zoos like San Diego, and Birmingham in Alabama under the guidance of CEO Chris Pfefferkorn. The European Elephant Management School at Hamburg's Tierpark Hagenbeck, led by Dr Stephan Hering-Hagenbeck at the time, was one of the first places in Europe to get keepers together collectively thinking about the future and what was needed for elephants. I'm also very inspired by the work of Lisa New at Knoxville Zoo.

I've admired Chester Zoo for decades, and its philosophy of in situ conservation in the home-range countries for many different species has been inspiring. It was supporting projects in the wild long before most other zoos. I always look to see what it's doing in elephant care, and what improvements it's proposing to its habitats, as it is often at the forefront of progress. Some of the best keepers I have met have come from Chester, including Alan Roocroft and Andy McKenzie, who has led their elephant programme for many years.

The modern zoo has become a conservation organisation that runs a zoo. Conservation cannot just be about handing over a cheque once a year. In addition to breeding more endangered species, it needs to become responsible for parts of the wild in Africa, India, Asia, South America and at home), supporting projects and the people who share their space with these fabulous species. We need to protect the animals and the rangers that protect them; help communities cohabitate with animals; and develop strategies to reintroduce endangered species.

If Howletts' rewilding plan for elephants in Africa succeeds, it is likely more zoos will follow their lead. There have already been successful reintroductions of zoo-bred species, including the scimitar-horned oryx and golden lion tamarin, but we need more animals going back to their natural habitats – from bugs to elephants, and everything in between, and not just the keystone species. Zoos also need to look at smaller invertebrates, birds and reptile species, both at home and abroad. The world's animals are under siege, and the reality is that in the next 50 years, hundreds and hundreds of species will be extinct.

Supporting the communities who live side by side with these species is key. It's very easy for me to say, 'Don't shoot an elephant or a tiger to sell the body parts' when I can walk into McDonald's and get fed. Some people in Indonesia, China and Malaysia live on the equivalent of $100 a year. Shooting a tiger might get you $50. If I had six kids in a hut, I'd kill a tiger to support my family – we all would. We need to help families and villages – by selling their crafts, as Dublin Zoo does, or by showing them new sources of revenue, buying them land or creating meaningful work.

We in the Western world also need to continue to review and improve how we keep elephants in human care. I love Dublin Zoo. I love what we have achieved for elephant welfare, but we need to keep re-evaluating and continue to plan for the future. The secret of good elephant care is constantly reviewing what we are doing.

Compromised welfare has to stop – there's been too much talking and not enough action. We have to stop using fear to manage elephants. We have to stop keeping them in sterile surroundings. We have to stop female elephants from being shackled for hours on end during labour.

The cost of improving elephant facilities cannot be an excuse any longer. Every zoo can make changes that will create mental and physical stimulation for their elephants. And if a facility cannot look after an elephant the way it should be cared for, they need to send them somewhere that can. Dublin Zoo now understands the shortcomings of how elephants were kept, and they have made it right. However, many zoos still keep elephants in small and inappropriate conditions.

Bodies like the EAZA and BIAZA are constantly reviewing standards, and zoos continue to strive to make habitats more spacious and dynamic. However, space is only good when it's quality space that can be utilised for appropriate elephant behaviour; where they have motivation, constant choice and opportunities for meaningful movement. The role of zoos in highlighting elephant conservation is very important, but we cannot keep elephants unless we provide them with activities that allow them to behave as they do in the wild: foraging, playing, sleeping, reproducing and socialising. We also need to agree that elephants should be in multigenerational herds. If those conditions are not mirrored in a zoo situation, we have no justification to keep them. The key to improving the welfare of elephants in human care is held by the elephants themselves. Their natural biology, physiology and behaviour should be the blueprint for any institution hoping to keep elephants in long-term human care.

There have to be global standards of elephant care and inspections. The EAZA has said that all its zoos need to use protected contact by 2030, but zoos need to be shown how and supported. What saddens me when I go to different zoos is to see the variation in care, and how everybody thinks they are doing it right. Currently, there are even different interpretations of what protected contact means. There should also be an international standard of elephant keeping and training – trained keepers should be able to travel from the US to Europe, Australia or China and fit into a team that provides the same consistency and quality of care. Unfortunately, we are a long way off that.

Our aspiration should be that all elephants in human care enjoy a similar standard of quality living, one that allows for mental and physical stimulation, social needs, and the opportunities and choices to live a meaningful life. Elephants need to be happy and healthy. They need life plans that are based on an understanding of their life cycles and what they need at critical times in their lives of 50 or 60 years.

Our attention should not just be on elephants, either. Good zoos are now keeping fewer species with more space. There are growing numbers of specialist zoos that keep only ape species or carnivores – smaller numbers of highly endangered species being given a better quality of life.

However, what animals we keep in zoos must be given habitats as close to the wild as possible. We have to harness technology and use every means possible to replicate how they would live in the wild – what Dr Jake Veasey from Care for the Rare would describe as, 'dynamic habitats and meaningful locomotion'. We have to keep pushing the boundaries in design, creating climatic zones, changing landscapes, providing opportunities for them throughout the day and night, keeping predators moving for days in search of food and letting swimmers swim and diggers dig. We are limited only by our imaginations.

There has been a lot of talk, and rightly so, about ape rights. But we have to put all animals in human care under the spotlight, including animals in the food chain. Everything needs to be scrutinised and questioned, even actions like lowering lobsters into boiling water when it has been proven that they feel pain.

We all have a role to play. Examine your conscience and support only ethical tourism. Animal magnetism can be hard to resist, but you need to realise that the temple tiger you are admiring has been drugged, and the elephant you are riding has been bashed into submission so you can sit on its back. When you watch a parrot show in Spain, remember that all these birds have likely been mutilated, their wings pinioned to stop them from flying away. The palm oil in your biscuits and soap probably comes from a forest where orangutans and tigers once lived. Let your supermarket know that you only want products that use sustainable palm oil. We are at the bottom of the hill in terms of the decline in species. There are 14,000 species on the brink of extinction. We all have our part to play to protect the natural world.

As a sit here, I have a fragile optimism for the future; an optimism that we can do better for elephants in human care. Dublin Zoo has been a beacon of hope, but all zoos now need to work together and charge into the future. Elephants have to be at the core of everything we do. They are a unique, wonderful species that have brought nothing but joy to humans. Now we need to bring joy back to their lives with meaningful conservation and strategies.

Kindness is the key to future success. It needs to be at the heart of how we manage elephants in human care, how we manage the human–animal conflict in what is left of their wild, and how we care for ourselves and the world around us.

ACKNOWLEDGEMENTS

I would like to thank my father for his guidance as a zookeeper and for giving me my start, and Alan Roocroft for mentoring me over the years. I would also like to thank Christoph Schwitzer for taking the time to review this manuscript before we went to print.

Finally, thank you to Louise Ní Chríodáin, Seán Hayes, Margaret Farrelly and the entire team at Gill for making this book possible.